Modes of Representation in Developing Statistical Thinking in Education

Sofia D. Anastasiadou
University of Western Macedonia, Greece

Lamprini Seremeti
Agricultural University of Athens, Greece

IGI Global
Scientific Publishing
Publishing Tomorrow's Research Today

Vice President of Editorial — Melissa Wagner
Director of Acquisitions — Mikaela Felty
Director of Book Development — Jocelynn Hessler
Production Manager — Mike Brehm
Cover Design — Jose Rosado

Published in the United States of America by
IGI Global Scientific Publishing
701 East Chocolate Avenue
Hershey, PA, 17033, USA
Tel: 717-533-8845
Fax: 717-533-7115
Website: https://www.igi-global.com E-mail: cust@igi-global.com

Library of Congress Cataloging-in-Publication Data

Library of Congress Cataloging-in-Publication Data

Names: Anastasiadou, Sofia D., 1968- editor | Seremeti, Lamprini, 1974- editor
Title: Modes of representation in developing statistical thinking in education / edited by: Sofia Anastasiadou, Lamprini Seremeti.
Description: Hershey, PA : IGI Global Scientific Publishing, [2025] | Includes bibliographical references and index. | Summary: "The primary objectives of the collective volume are to showcase the latest advancements in didactical techniques, highlight innovative didactical techniques across fields of statistics, probabilities and mathematics, and provide a comprehensive resource for understanding and addressing the complexities of learning statistical notions. By bringing together contributions from global experts, the volume aims to foster interdisciplinary collaboration, inspire new research directions, and equip readers with the knowledge and tools necessary to navigate the evolving landscape of statistics education"-- Provided by publisher.
Identifiers: LCCN 2025001356 (print) | LCCN 2025001357 (ebook) | ISBN 9798369399347 hardcover | ISBN 9798369399354 paperback | ISBN 9798369399361 ebook
Subjects: LCSH: Statistics--Study and teaching
Classification: LCC QA276.18 .M63 2025 (print) | LCC QA276.18 (ebook) | DDC 519.5071--dc23/eng/20250428
LC record available at https://lccn.loc.gov/2025001356
LC ebook record available at https://lccn.loc.gov/2025001357

British Cataloguing in Publication Data
A Cataloguing in Publication record for this book is available from the British Library.

Table of Contents

Detailed Table of Contents

Statistical thinking is not just about performing mathematical computations but about interpreting and making sense of data to draw conclusions and inform decisions. In educational contexts, this means students need to understand how to collect, analyze, and evaluate data, as well as how to communicate their findings effectively. While individual practice is important, the role of group learning in developing statistical thinking cannot be overstated. Group learning encourages collaboration, allows students to engage with diverse perspectives, and helps them overcome cognitive biases that might arise when working alone. By working in groups, students have the opportunity to discuss statistical concepts, apply them to real-world situations, and deepen their understanding through peer interactions. This chapter presents a novice approach to group-based personalized learning, leveraging machine learning (ML) - driven clustering algorithms for dynamic group formation.

Artificial Intelligence (AI) is revolutionizing multiple fields, including education. In the domain of statistics, AI presents new opportunities for enhancing the development of statistical thinking by offering advanced modes of representation. These AI-driven tools can support learners in understanding complex statistical concepts through visualization, interactive simulations, automated feedback, and personalized learning environments. This chapter explores how AI can be effectively integrated into statistical education to foster deeper engagement, understanding, and application of statistical thinking. By discussing case studies, practical applications, and the potential challenges, this chapter outlines a roadmap for using AI to transform statistical education.

The present study explores the similarities and differences between two distinct statistical methods: Principal Component Analysis (PCA) and Implicative Statistical Analysis. Specifically, the study assesses the results derived from applying these two methodologies to a case study on Greek students' understanding of probability distributions (e.g., Binomial Distribution, Poisson Distribution, Normal Distribution). The research involved 119 second-year students from the Department of Educational and Social Policy at the University of Macedonia. The primary objective was to determine whether these two statistical approaches yield similar or different conclusions regarding student comprehension. Additionally, the specific advantages of each statistical method are discussed in detail.

This chapter shows the development, implementation, & evaluation of comprehensive desktop-based software tool engineering using Python to facilitate an array of set operations. Software furnishes users with a versatile platform for conducting fundamental set manipulations (e.g., union, intersection, computation, difference, symmetric different, power set generation, & Cartesian product computation). It features a user-friendly interface with input fields for two sets with clearly labelled buttons for each operation. This design aims to make the software easy to use for everyone, from beginners to experts, enhancing accessibility and user satisfaction. Through judicious utilization of python functions skillfully designed to capture the complexities of each set operation, it ensures seamless execution and expedient delivery of precise results in real-time. Algorithms supporting each operation are thoroughly explained, clarifying underlying computational methods & theoretical foundations. Its effectiveness & usefulness are rigorously assessed.

Statistics, often referred to as data science, is a powerful tool for understanding the world. It allows us to identify patterns, test hypotheses and make predictions in fields as diverse as medicine, economics, social sciences and technology. However, the power of statistics lies not only in its numbers, but also in its semantics - the meaning and interpretation we derive from them. Understanding the semantics of statistics requires us to identify the relationships between data, context and narrative, recognizing both the opportunities for insight and the risks of misinterpretation. In this vein, the purpose of this chapter is twofold; first, to clarify the difference between "statistical semantics" and "semantics of statistics", and second, to explore the role of ontologies, as formal frameworks to represent knowledge, in the semantics of statistics.

This chapter investigates the impact of data visualization on enhancing statistical thinking in higher education, emphasizing its significance in developing students' abilities to analyze and interpret data effectively. By utilizing visual tools such as graphs, charts, and interactive simulations, students can transform abstract statistical concepts into concrete understandings, facilitating deeper engagement and comprehension. The study explores various technologies, including R, Python, and Tableau, that support dynamic data manipulation and real-time feedback, promoting active learning experiences. While the benefits of visualization are substantial, challenges such as varying levels of technological proficiency among students and the potential for superficial understanding are also addressed. Ultimately, this research underscores the importance of integrating effective visualization techniques with pedagogical strategies to cultivate robust statistical literacy and critical thinking skills, preparing students for data-driven decision-making in their respective fields.

Courses such as probability cause anxiety in students even today. Anxiety along with other negative feelings such as fear and anxiety hinder success in learning. They additionally hinder plans for future engagement with the subject of probabilities. In the present research we study students' attitudes about probabilities. 162 students, of which 69 were female students and 93 male students completed a questionnaire assessing their attitudes towards the course. The results showed that students have very negative attitudes connected with the anxiety they feel about statistics and probability, and they consider that although they are very useful subjects and they consider that these subjects are very useful for research.

This paper investigates the effectiveness of teaching different kinds of representations in teaching probabilities. Verbal, algebraic and graphical representations are used in teaching Venn diagrams. A sample of 127 students was used to examine their effectiveness in teaching and learning probability concepts. Implicative statistical analysis is applied to evaluate the relation between students' approach and their ability to solve problems. More exactly, this study investigates 127 students' abilities in using representations (verbal, tabular, graphical and symbolic form) of basic statistical concepts and in moving from one mode of representation to another successfully. Diagrammatic modes seem to be easier than the other nodes and are linked to higher performance.

Preface

INTRODUCTION

Statistical thinking lies at the heart of data-driven decision-making, a cornerstone in both academic disciplines and real-world applications. As education evolves, the challenge is not merely to teach statistical procedures but to nurture learners who can critically analyze, interpret, and communicate data insights. This edited volume, *Modes of Representation in Developing Statistical Thinking in Education*, responds to this need by exploring how varied representations—visual, symbolic, contextual, and technological—can deepen students' statistical understanding.

The increasing integration of technologies like artificial intelligence, machine learning, and data visualization into educational contexts introduces new possibilities for enriching statistical instruction. These tools, when combined with cognitive strategies and representational diversity, provide learners with more intuitive and meaningful ways to engage with abstract concepts. This book offers a comprehensive look at how such integration can be implemented across secondary and higher education settings.

Moreover, this volume brings together empirical studies, theoretical reflections, and practical tools aimed at addressing common student misconceptions and emotional barriers related to statistics. Through the lens of representation and pedagogy, contributors present strategies that bridge the gap between theory and practice, enhancing statistical literacy and cognitive engagement for diverse student populations.

CHAPTER OVERVIEW

Chapter 1: Enhancing Statistical Thinking Through ML-Driven Clustering Algorithms for Group-Based Learning

This chapter introduces a personalized learning approach that uses machine learning-driven clustering algorithms to form dynamic student groups. It emphasizes how collaboration enhances statistical understanding by encouraging dialogue, critical thinking, and mutual problem-solving.

Chapter 2: Integrating AI Into Statistical Education: New Modes of Representation for Developing Statistical Thinking

Exploring the transformative potential of artificial intelligence in education, this chapter outlines how AI-based visualization, interactive simulations, and feedback tools can be employed to support the development of statistical thinking and personalized learning environments.

Chapter 3: Principal Component Analysis vs. Implicative Statistical Analysis: The Case of Probability Distributions' Understanding

This comparative study evaluates two statistical methods—Principal Component Analysis and Implicative Statistical Analysis—in assessing students' grasp of probability distributions, highlighting the unique insights each method provides into learning outcomes.

Chapter 4: Python-Powered Set Operation—An Interactive Web Tool for Discrete Mathematics Education and Application

A hands-on chapter showcasing a Python-based educational software designed to support set operations. The tool's functionality, user interface, and underlying algorithms are discussed, illustrating its application in teaching mathematical and statistical concepts.

Chapter 5: Representations in Developing Statistical Thinking in Secondary Education

This chapter emphasizes the role of representations as cognitive tools to translate abstract statistical ideas into tangible forms. It discusses how visual and symbolic representations foster conceptual clarity and application in real-life contexts for secondary education students.

Chapter 6: Representations in Developing Thinking Regarding Probabilities in Higher Education

Focusing on the challenges of teaching probability, this chapter explores how multiple representations—visual, symbolic, and interactive—support higher education students in understanding complex probability concepts and applying them confidently.

Chapter 7: Students' Misconceptions in Learning Probability Distributions

By analyzing student responses and applying implicative statistical analysis, this chapter identifies common misconceptions in understanding various probability distributions and offers pedagogical strategies to address these errors.

Chapter 8: The Role of Multiple Representations in Enhancing Statistical Thinking in Secondary Education: A Case Study Approach

Through qualitative case studies, this chapter illustrates how the use of graphical, numerical, and verbal representations enhances students' cognitive processing and data interpretation, while also addressing the challenges in transitioning across modes.

Chapter 9: The Semantics of Statistics

This chapter delves into the interpretative aspect of statistics, distinguishing between statistical semantics and the semantics of statistics. It examines the use of ontologies and formal frameworks to represent meaning in data and avoid misinterpretation.

Chapter 10: Visualizing Insight: Enhancing Statistical Thinking in Higher Education

Data visualization is explored as a catalyst for statistical insight. This chapter examines how tools like R, Python, and Tableau transform student learning experiences by providing real-time, interactive feedback and making abstract concepts more accessible.

Chapter 11: Students' Attitudes Toward Statistics and Probabilities

This empirical study investigates the emotional landscape of statistics learners, highlighting how anxiety and negative attitudes affect engagement. It emphasizes the importance of addressing affective barriers in statistical instruction.

Chapter 12: The Role of Different Modes of Representation in Statistics

Analyzing the efficacy of verbal, algebraic, and graphical representations in teaching Venn diagrams, this chapter explores students' ability to transition across representational forms and how this ability correlates with improved performance in probability tasks.

CONCLUSION

The chapters in this volume collectively demonstrate that the development of statistical thinking is deeply intertwined with how content is represented and communicated. By employing a wide range of representational tools—from AI simulations to diagrammatic reasoning—educators can foster richer, more meaningful learning experiences. These representations act not only as instructional aids but as bridges between abstract theory and practical understanding.

An essential insight emerging from this collection is the need for intentional and varied use of representations in teaching, especially in areas that are abstract or prone to student misconceptions. Whether addressing emotional barriers, clarifying semantic interpretations, or guiding students through complex analytical tasks, representation functions as a foundational pedagogical strategy.

This book invites educators, researchers, and curriculum designers to rethink traditional approaches and embrace more dynamic, inclusive, and technologically enriched methodologies. As we navigate a data-saturated world, equipping students with strong statistical thinking through diverse representational modes becomes not only beneficial but imperative.

Sofia D. Anastasiadou
University of Western Macedonia, Greece

Lamprini Seremeti
Agricultural University of Athens, Greece

Chapter 1
Enhancing Statistical Thinking Through ML-Driven Clustering Algorithms for Group-Based Learning

Charis Kalogera
Agricultural University of Athens, Greece

Lambrini Seremeti
https://orcid.org/0000-0002-0663-5408
University of Western Macedonia, Greece

ABSTRACT

Statistical thinking is not just about performing mathematical computations but about interpreting and making sense of data to draw conclusions and inform decisions. In educational contexts, this means students need to understand how to collect, analyze, and evaluate data, as well as how to communicate their findings effectively. While individual practice is important, the role of group learning in developing statistical thinking cannot be overstated. Group learning encourages collaboration, allows students to engage with diverse perspectives, and helps them overcome cognitive biases that might arise when working alone. By working in groups, students have the opportunity to discuss statistical concepts, apply them to real-world situations, and deepen their understanding through peer interactions. This chapter presents a novice approach to group-based personalized learning, leveraging machine learning (ML) - driven clustering algorithms for dynamic group formation.

DOI: 10.4018/979-8-3693-9934-7.ch001

1. INTRODUCTION

In today's data-driven world, statistical thinking is a crucial skill that extends beyond the mere ability to perform mathematical computations (Loukat et al., 2008; Chance, 2017). It involves interpreting data, drawing meaningful conclusions, and making informed decisions, skills that are essential across numerous domains (Souravlas et al., 2023; Souravlas & Anastasiadou, 2022; Souravlas et al., 2022). In educational settings, fostering statistical thinking goes beyond teaching technical procedures (Anastasiadou, 2005; 2008; 2010; Anastasiadou et al. 2008). It requires students to understand how to collect, analyze, and evaluate data, and how to effectively communicate their findings using different modes of representations (Anastasiadou & Loukas, 2009; Anastasiadou, 2007; 2019a; 2019b). While individual practice plays a role in this development, group learning has emerged as an important strategy in enhancing students' understanding of statistical concepts. Group learning promotes collaboration, allows students to engage with diverse perspectives, and aids in overcoming cognitive biases that can occur when working in isolation (Gunderson and Moore, 2008). When students work together, they can discuss and apply statistical methods to real-world problems, deepening their understanding through peer interactions.

In this vein, this chapter presents a novice approach to personalized learning that shifts from traditional individual-focused strategies to group-based personalized learning, with a particular emphasis on enhancing students' statistical thinking. The originality of this approach lies in forming instantaneous student groups tailored to shared learning needs, abilities, and interests, using a machine learning (ML)-driven clustering algorithm. This approach is particularly valuable in the context of statistical education, where students often face challenges in grasping abstract concepts, interpreting data, and reasoning critically about statistical problems.

In particular, in the context of statistical thinking, instantaneous group formation through clustering algorithms offers several advantages. Statistical thinking involves interpreting data, making sense of patterns, and drawing conclusions from real-world situations, skills that can be better developed in a collaborative environment. By forming groups on the fly, students can work with peers who complement their strengths and weaknesses, fostering an environment where they can challenge each other's assumptions, address misunderstandings, and refine their reasoning. Clustering algorithms can account for a range of student characteristics, such as prior knowledge of statistical concepts, problem-solving approaches, and even emotional readiness to tackle complex topics (Pang et al., 2014). This ensures that students are grouped in ways that maximize their learning potential, facilitating peer learning and deeper engagement with the subject matter. Moreover, the flexibility of clustering algorithms makes it possible to adjust groupings as students progress, ensuring that group dynamics remain conducive to learning throughout the course, particularly as statistical problems evolve and become more challenging.

2. THE LEARNING CLUSTERING ALGORITHM

This section explores the concept of using hierarchical clustering methods (Bindra and Mishra, 2023), a well-known technique in artificial intelligence, and in particular in ML, to form learning groups in the classroom. Hierarchical clustering groups data points based on their similarity, making it an ideal choice for clustering students according to various attributes such as academic performance, participation, cooperation, and personal interest. The process begins with a set of students who are assigned attributes based on their academic assessments, behavioral data, and teacher feedback. These attributes are then transformed into numerical values, ensuring the data is normalized for effective analysis.

The clustering algorithm relies on a distance calculation method (such as Euclidean or Manhattan distance) to measure similarities (Gao et al., 2023) between students. Through this iterative process, the algorithm forms groups based on the proximity of the students' attributes, resulting in a hierarchical tree that visually represents the relationships between them. This dynamic grouping allows students to collaborate with others who complement their strengths and weaknesses, providing a diverse learning environment that facilitates deeper engagement with statistical thinking.

• Describing the Algorithm

Instantaneous grouping within classroom can be achieved through hierarchical clustering, an algorithm used in data analysis and machine learning to group data based on similarity or distance between them. The implementation of such an algorithm requires programming skills (e.g. Python with libraries such as NumPy, Pandas and SciPy) together with skills in statistics and data analysis.

The algorithm starts with a set of students who have been assigned attributes. These attributes are derived from the students' ratings and the portfolio data. The data collection includes assessments in various courses, as well as attributes provided by teachers, such as participation, interest, and cooperation. In addition, performance data from previous years and other relevant information is collected.

Data preparation and analysis involves transforming and mapping categories into numerical values such as scores and percentages. Normalization of the data is necessary to ensure that different scales do not affect the clustering process. Entities (students) start as independent, self-contained units. As the clustering progresses, the algorithm focuses on one or more characteristics to classify the students. A distance calculation method (e.g., Euclidean distance, Manhattan distance) is applied to measure the similarity between students. Hierarchical clustering is used to create a tree that depicts the relationships between students based on the distances of the features being studied.

Clustering begins by measuring the “distance” between entities based on predefined criteria. Therefore, groups are distinguished solely by the characteristics that led to their classification. The main attribute used in the distance calculation determines the classification of the groups. A distance matrix is then created and updated during the clustering process.

• Hard vs. Soft grouping

Clustering can be classified into hard and soft grouping techniques (Reddy and Vinzamuri, 2018). In hard clustering, each student belongs strictly to one group, defined by a specific characteristic (e.g., students excelling in Statistics computation). Soft clustering, on the other hand, allows for multiple group memberships, where the degree of belonging to a group depends on the intensity of the characteristics being measured. For example, a student may belong to both a Statistics group that can easily calculate statistical measures and another Statistics group that can easily interpret statistical data, depending on their performance in this subject. This flexibility ensures that each student's strengths are recognized, promoting a more inclusive and personalized learning experience.

• Visualization of the Grouping

We start with a set of entities (students), which initially form independent groups (Figure 1).

Figure 1. Total number of students

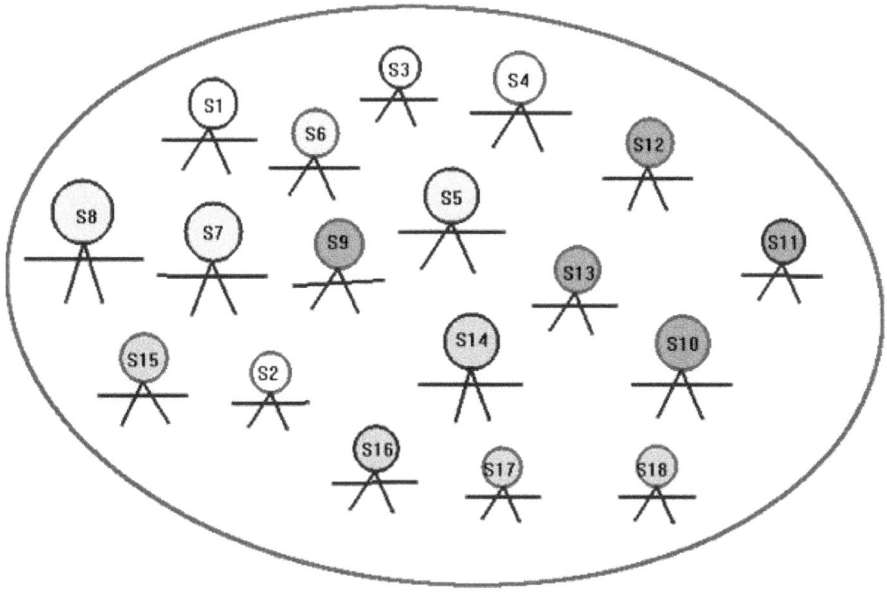

We then create a distance matrix using the metric we choose based on the feature we want to study. For example, Table 1 might focus on students' behavior in Statistics calculation (feature-SC) or Statistics interpretation (feature-SI), where DSC(si, sj) = min{ d(fsi, fsj) } i= 1,2,3,……

Table 1. Matrix recording the distances based on characteristic SC

	s1	**s2**	**s3**	**s4**	**…..**	**s18**
s1						
s2		Dsc(s2, s2)				Dsc(s2,s18)
s3						
. . .						
s18		Dsc(s18,s2)				

The matrices are based on the chosen characteristics, and the hierarchical clustering algorithm proceeds to create a tree-like diagram showing the relationships between students.

Figure 2. A tree diagram illustrating the step-by-step merging of students into groups based on calculated distances

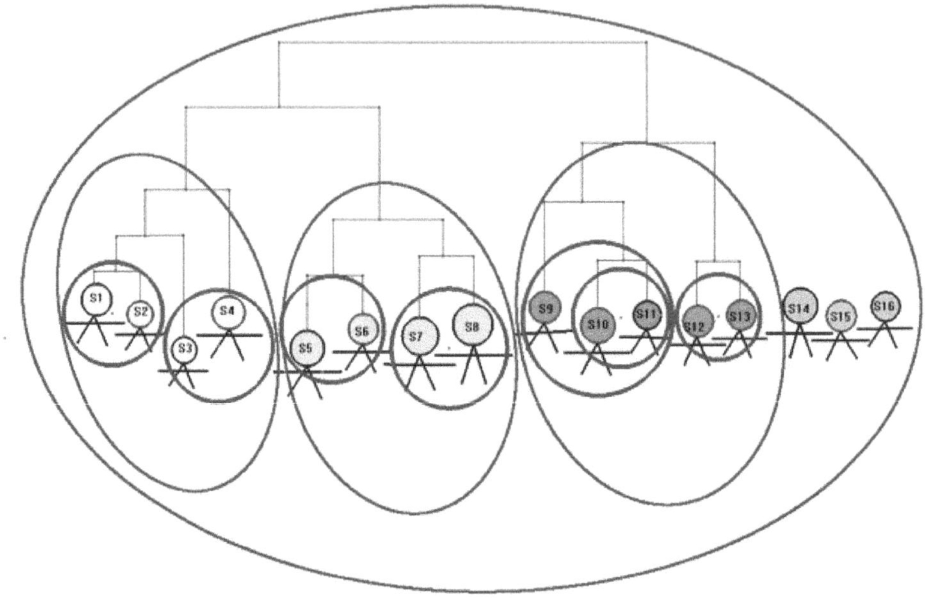

The algorithm merges the closest groups based on their calculated distances, iterating until a complete hierarchical structure is created.

Figure 3. The closest groups based on their calculated distances

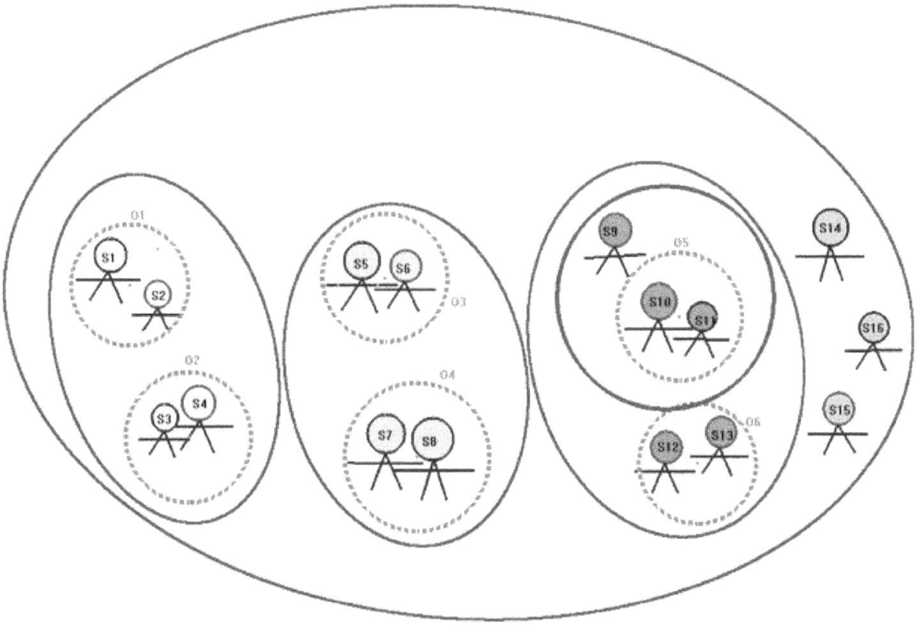

• The Hierarchical Tree Diagram

The hierarchical tree diagram illustrates the creation of groups. After several steps of comparing entities and measuring distances, the distance matrix is updated. Then, entities with the shortest distance between them are merged, resulting in a hierarchical grouping structure (Figure 3).

Figure 4. The tree diagram showing the groups being created

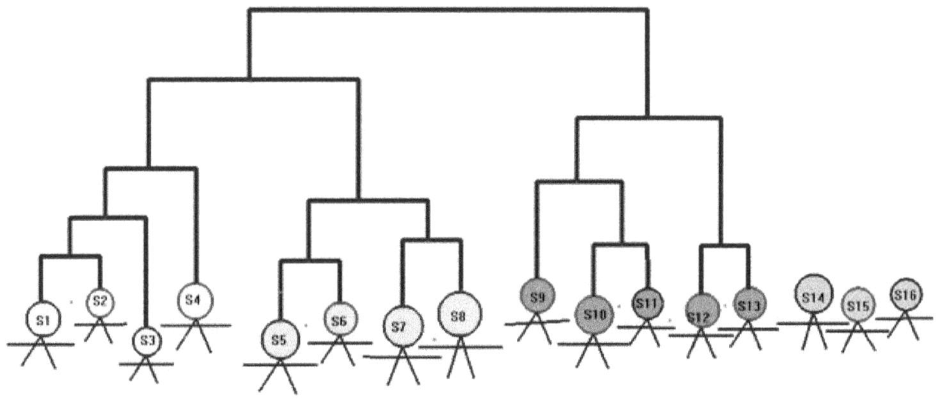

At each step, the distance matrix is updated, and entities with the smallest distance are merged to form a larger group. The process continues until all students are part of one final group.

- Grouping Using Distance

One common method for hierarchical clustering is the single-link distance, which calculates the shortest distance between any pair of points in two groups. That is, we create the groups according to the distances of the entities such as O1 of s1, s2 and O2 of s3, s4. We repeat the process until the set of groups creates the class of students. Single-link distance between groups Oi and Oj is the shortest distance between all points of group Oi and group Oj. The distance is defined by the mathematical formula Ds(Oi, Oj) = mind(x,y){ d(x, y)|xeOi, yeOj }.

In hierarchical clustering the distance of groups is defined between two points of entities created by the first measurement that have the minimum distance between them. In the case of groups O1 and O2, the distances s1-s3, s1-s4, s2-s3, s3-s4 will be calculated to find the minimum distance and join O1 and O2 to create a larger group with students of similar characteristics.

Table 2. Simple Union distance matrix – an example

	O1	**O2**	**O3**	**O4**	**O5**
O1	1,00	0,95	0,05	0,67	0,22
O2	0,95	1.00	0,70	0,62	0,50
O3	0,05	0,70	1,00	0,38	0,30
O4	0,67	0,62	0,38	1,00	0,85
C55	0,22	0,50	0,30	0,85	1,00

The distance between two groups has many alternative measurements and depends on the characteristics of the group of students that we define for measuring the distance. Each group is a set of students. Determining the points for measuring the distance depends on the method and the clustering algorithm we follow. There are many alternatives and it is a complex process.

The distance is determined by mathematical formulas but the result is measured in quantities depending on the characteristics we assign to the student entities.

• Functioning of the Algorithm

The algorithm starts with a set of students, each assigned attributes derived from academic performance, teacher feedback, and participation metrics. These attributes are transformed into numerical values, such as scores and percentages, and normalized to ensure consistency across different scales. Initially, each student is treated as an independent entity (Figure 1). The algorithm calculates the "distance" between students based on their attributes. This distance represents how similar or different the students are. A distance matrix is generated, where each cell represents the measured distance between two students (Table 1 and 2). The algorithm proceeds by grouping the students with the shortest distance (most similar characteristics) into clusters. It updates the distance matrix after each grouping step (Table 3) and continues until all students are merged into a single hierarchy. The algorithm uses specific methods, such as single-link or complete-link distances, to refine group boundaries. This ensures that groups reflect meaningful similarities based on the selected characteristics. After clustering, the resulting groups are visualized. These groups can be represented using diagrams that illustrate their composition and relationships. For example, flexible nested arrangements can show the overlapping or distinct nature of groups.

Table 3. Table of updated distances in the groups

	O1	**O2**	**O3**	**O4**	**O5**
O1					
O2					
O3					
O4					(O4)U(O5)
O5				(O5)U(O4)	

The algorithm merges the two closest groups (e.g., O4 and O5) and updates the distance matrix. This process repeats until all students are grouped into a single final structure (Figure 4).

Figure 5. The final structure

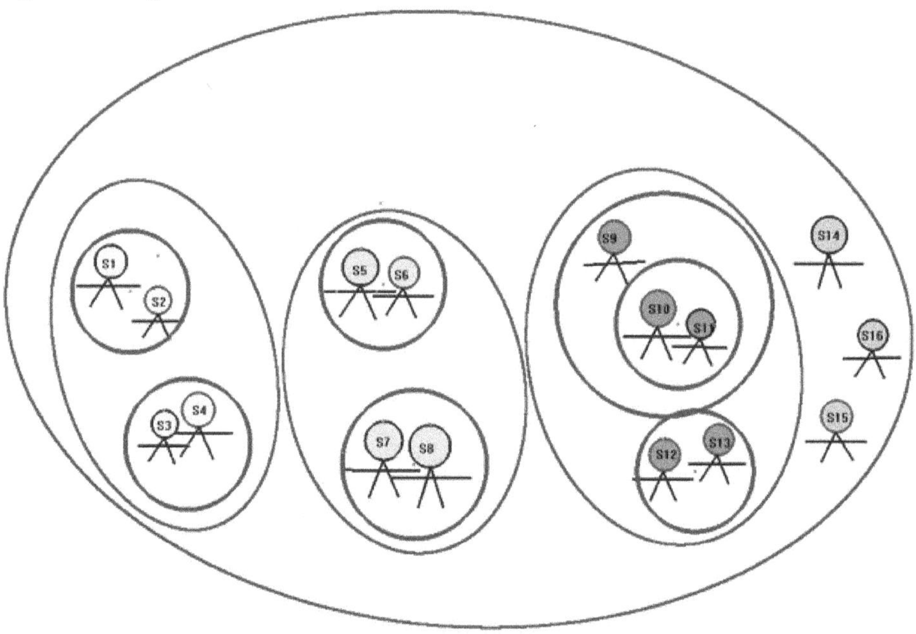

The grouping described can be observed in every learning classroom. Its significance lies in supporting students based on the interconnected group they belong to (Kougias et al., 2012). Additionally, by linking the grouping with learning representations, a comprehensive learning environment is created. Recognizing the groups contributes to the development of a democratic teaching approach rather than segregating students into different ability levels (Kougias et al., 2016).

The applications and benefits are expected to contribute to personalized teaching, allowing educators to tailor their instruction according to the needs of the groups. Students can collaborate with others who have similar characteristics and learning needs. The analysis of groups can provide valuable insights into the progress of students.

3. CONCLUSION

Group-based learning is an essential tool for enhancing statistical thinking in the classroom. By working with their peers, students are exposed to diverse ideas, challenged to think critically, and encouraged to engage deeply with statistical concepts. Group-based learning not only supports the development of statistical reasoning but also promotes key skills such as problem-solving, communication, and peer teaching. The application of clustering algorithms in statistics education can further enhance group learning by ensuring that students are placed in optimal groups based on their performance, learning preferences, and social dynamics. This approach promotes a more group-based personalized and effective learning experience, helping students to strengthen their statistical thinking and apply it to real-world challenges.

REFERENCES

Anastasiadou, S. (2005). Affective reactions and attitudes of the last class of Greek high school students towards statistics, In M Boosch (Eds) *Proceedings of the Fourth Conference of the European Society for Research in Mathematics Education* (CERME 4), Sant Feliu de Guíxols, Spain.

Anastasiadou, S. (2007). T*he Role of representations in solving statistical problems and the translation ability of fifth and sixth grade students.Proceedings of the International Journal of Learning*, *14*(10), 125–132.

Anastasiadou, S. (2008). Iconic Representations and Statistical Problem Solving: The Case of Greek Pupils. *International Journal of Learning*, *15*(11), 301–310. DOI: 10.18848/1447-9494/CGP/v15i11/58699

Anastasiadou, S. (2010). *Pre-service teachers' performance on the learning of probability distributions and the role of projects: A multilevel statistical analysis.* In Regnier, Jean-Claude, Spagnolo, Filippo, Di Paola, Benedetto, & Gras, Regis, (Eds), *Proceedings of the 5th International Conference Implicative Statistic Analysis*, Palermo, Italy pp 73-84.

Anastasiadou, S. (2019a). *Representations in statistics problem solving.* ICERI2019, Proceedings of the 12th annual International Conference of Education, Research and Innovation, Seville (Spain), ICERI2019, pp. 8993- 8996. DOI: 10.21125/iceri.2019.2159

Anastasiadou, S. (2019b). *Internal representations toward external representations: The case of statistical concepts.* 11th annual International Conference on Education and New Learning Technologies, Palma de Mallorca, Spain, EDULEARN 19, pp. 6-9. DOI: 10.21125/edulearn.2019.0004

Anastasiadou, S., Elia, I., & Monoyiou, A. (2008). The effects of representations on the learning and understanding of statistics in Greek primary school. *Mediterranean Journal for Research in Mathematics Education*, *7*(1), 67–80.

Anastasiadou, S., & Loukas, D. (2009). Greek pre-service teachers' cognitive abilities in understanding the concept of frequency: A multilevel statistical analysis. *International Journal of Learning*, *16*(5), 189–202. DOI: 10.18848/1447-9494/CGP/v16i05/46268

Bindra, K., & Mishra, A. (2023). A detailed study of clustering algorithms. *Proceedings of the 6th International Conference on Reliability: Infocom Technologies and Optimization (Trends, and Future Directions)*, Noida, India, 371-376. DOI: 10.1109/ICRITO.2017.8342454

Chance, B. L. (2017). Components of statistical thinking and implications for instruction and assessment. *Journal of Statistics Education : An International Journal on the Teaching and Learning of Statistics, 10*(3). Advance online publication. DOI: 10.1080/10691898.2002.11910677

Gao, C. X., Dwyer, D., Zhu, Y., Smith, C. L., Du, L., Filia, K. M., Bayer, J., Menssink, J. M., Wang, T., Bergmeir, C., Wood, S., & Cotton, S. M. (2023). An overview of clustering methods with guidelines for application in mental health research. *Psychiatry Research, 327*, 115265. DOI: 10.1016/j.psychres.2023.115265 PMID: 37348404

Gunderson, D., & Moore, J. O. (2008). Group learning pedagogy and group selection. *International Journal of Construction Education and Research, 4*(1), 34–45. DOI: 10.1080/15578770801943893

Kougias, I., Seremeti, L., & Kalogeras, D. (2012). Information flow among dynamic and distributed educational environments: A formal framework. *Proceedings of the Innovative Approaches in Education – Design and Networking Conference*, 164-172.

Kougias, I., Seremeti, L., & Kalogeras, D. (2016). An innovative resilient strategy towards creatively teaching mathematics. *International Journal for Talent Development and Creativity, 4*, 155–163.

Loukas, D., Papadimitriou, I., Drosos, G., & Anastasiadou, S. (2008). *Investigation of a categorical data distribution via the methods of multivariate data analysis – An application to a stochastic multicriteria group decision-making problem.* In Risto Lahdelma, Kaisa Mierrinen, Pekka Salminren, Ahti Salo (Eds). Proceedings of the 67th Meeting of the European Working Group "Multiple Criteria Decision Aiding", Rovaniemi, Lapland, Finland, pp. 151-162.

Pang, Y., Xiao, F., Wang, H., & Xue, X. (2014). A clustering-based grouping model for enhancing collaborative learning. *Proceedings of the 13th International Conference on Machine Learning and Applications*, Detroit, MI, USA, 562-567. DOI: 10.1109/ICMLA.2014.94

Reddy, C.K. and Vinzamuri, B. (2018). A survey of partitional and hierarchical clustering algorithms. *Data Clustering*, 87-110.

Souravlas, S., & Anastasiadou, S. (2022). On Implementing Social Community Clouds Based on Markov Models. *IEEE Transactions on Computational Social Systems*. Advance online publication. DOI: 10.1109/TCSS.2022.3213273

Souravlas, S., Anastasiadou, S., Economides, Th., & Katsavounis, S. (2023). Probabilistic Community Detection in Social Networks. *IEEE Access : Practical Innovations, Open Solutions*, *11*, 25629–25641. DOI: 10.1109/ACCESS.2023.3257021

Souravlas, S., Anastasiadou, S. D., Tantalaki, N., & Katsavounis, S. (2022). A Fair, Dynamic Load Balanced Task Distribution Strategy for Heterogeneous Cloud Platforms Based on Markov Process Modeling. *IEEE Access : Practical Innovations, Open Solutions*, *10*, 26149–26162. DOI: 10.1109/ACCESS.2022.3157435

Chapter 2
Integrating AI Into Statistical Education:
New Modes of Representation for Developing Statistical Thinking

Lambrini Seremeti
https://orcid.org/0000-0002-0663-5408
University of Western Macedonia, Greece

Lazaros Anastasiadis
University of Crete, Greece

ABSTRACT

Artificial Intelligence (AI) is revolutionizing multiple fields, including education. In the domain of statistics, AI presents new opportunities for enhancing the development of statistical thinking by offering advanced modes of representation. These AI-driven tools can support learners in understanding complex statistical concepts through visualization, interactive simulations, automated feedback, and personalized learning environments. This chapter explores how AI can be effectively integrated into statistical education to foster deeper engagement, understanding, and application of statistical thinking. By discussing case studies, practical applications, and the potential challenges, this chapter outlines a roadmap for using AI to transform statistical education.

DOI: 10.4018/979-8-3693-9934-7.ch002

1. INTRODUCTION

Statistical thinking is a cornerstone of modern decision-making, enabling individuals to interpret data, identify trends, and draw meaningful conclusions in an increasingly data-driven world (Souravlas et al., 2021), (Gonzalez et al., 2020). Traditional approaches to teaching statistics have typically relied on textbooks, static visualizations, and structured problem-solving exercises (Hasim et al., 2024). While these methods have been effective to some extent, they often struggle to engage students fully or to convey the dynamic nature of statistical concepts such as variability, correlation, and probability. As a result, learners may develop a superficial understanding that falls short of fostering the critical thinking and analytical skills required for practical application.

The rapid advancement of artificial intelligence (AI) presents an opportunity to transform statistical education by redefining how statistical thinking is taught and learned (Calonge et al., 2023). AI-powered tools have the capacity to go beyond static and linear teaching methods, offering dynamic and interactive modes of representation that cater to diverse learning styles. These tools enable students to manipulate data in real time, engage with complex simulations, and visualize abstract concepts in intuitive ways. Furthermore, AI systems can provide personalized feedback and adaptive learning pathways that respond to individual progress, helping students overcome challenges and deepen their understanding of key principles (Rane et al., 2024).

This chapter examines the potential of AI to revolutionize the teaching and learning of statistics by introducing innovative modes of representation. It highlights how AI-driven tools can make abstract concepts more accessible, promote active learning, and cultivate statistical reasoning skills. Through the discussion of practical applications, real-world case studies, and the challenges inherent in integrating AI into education, this chapter outlines a roadmap for harnessing the power of AI to develop statistical thinking in learners more effectively and engagingly than ever before.

2. THE TRANSFORMATIVE POWER OF AI IN STATISTICAL EDUCATION

AI has significantly influenced education, particularly through adaptive learning, intelligent tutoring systems, and data-driven decision-making tools (Zhai et al., 2021), (Mittal et al., 2024). In the realm of statistics education, these AI-driven technologies are transforming how concepts are represented, making abstract ideas more tangible and accessible (Morales-Vargas and Benitez-Medina, 2024). For instance, interactive data visualizations powered by AI go beyond static representations, enabling learners to manipulate data in real-time. Students can explore how changes in variables affect outcomes, such as adjusting parameters like mean and standard deviation to observe immediate shifts in probability distribution curves. This dynamic approach allows learners to grasp complex principles more intuitively.

Additionally, AI-driven simulations offer another avenue for enhancing statistical education by generating scenarios that help students understand the behavior of statistical models under various conditions (Souravlas et al., 2023). For example, learners can investigate how different sample sizes influence estimation accuracy or explore interactions between variables in regression models. These simulations create a hands-on learning environment that bridges theoretical knowledge with practical application.

AI's capabilities in natural language processing (NLP) further enrich the educational experience by providing immediate and tailored feedback (Shaik et al., 2022). Tools leveraging NLP can analyze written or spoken input from students and offer corrective guidance or detailed explanations to improve their understanding of statistical results. Moreover, personalized learning pathways facilitated by AI help students engage with content that adapts to their unique strengths and weaknesses. By analyzing a learner's progress, AI systems recommend specific problems, simulations, or resources tailored to their needs, fostering a deeper and more effective engagement with statistical concepts.

Real-world applications illustrate the transformative power of AI in statistical education (Ellis and Slade, 2023), (Quinones et al., 2024). Platforms like DataCamp utilize AI algorithms to personalize learning experiences, analyzing student progress and tailoring course recommendations accordingly. Open-source tools like CODAP integrate AI to enhance data literacy by allowing students to manipulate datasets, visualize results, and receive automated feedback. Furthermore, AI-driven assessment platforms, such as Gradescope, provide instant and meaningful feedback on open-ended statistical problems, addressing misconceptions and improving learning outcomes. Collectively, these technologies demonstrate how AI can revolutionize the teaching and learning of statistics, making it more interactive, personalized, and effective.

3. AI ENHANCES STATISTICAL REPRESENTATION

Modes of representation play a critical role in how statistical concepts are presented and understood by students. Traditionally, these modes include visual, numerical, and algebraic representations, each offering unique ways to engage with statistical ideas (Gilbert, 2010). AI has the potential to enhance these traditional approaches by introducing dynamic, interactive experiences that foster a deeper and more intuitive conceptual understanding (Alcover et al., 2024).

In visual representation, AI empowers students to engage with data dynamically, enabling real-time manipulation of variables and immediate observation of how changes affect graphs and charts. This interaction transforms abstract statistical concepts like correlation, variability, and probability distributions into tangible experiences. By actively exploring visualizations, students can develop a more robust understanding of relationships within data, enhancing their ability to interpret and analyze statistical outcomes.

Numerical representation benefits from AI's ability to simulate extensive datasets, offering students opportunities to practice statistical analysis in a variety of realistic contexts. This hands-on approach helps students build numerical fluency by exposing them to practical applications of statistical computations, making the learning process more engaging and relevant to real-world scenarios. Working with diverse datasets allows students to explore statistical patterns, anomalies, and insights that are essential for informed decision-making.

In algebraic representation, AI serves as an intelligent tutor, guiding students through the mathematical foundations of statistical concepts. Through step-by-step problem-solving support, AI tools can demystify complex processes such as deriving the equation of a regression line or understanding its implications. This personalized assistance helps students bridge the gap between theoretical knowledge and practical application, building their confidence in tackling mathematical challenges within statistics. By enhancing all these modes of representation, AI not only enriches the learning experience but also equips students with the skills and understanding needed to excel in statistical reasoning.

4. CHALLENGES OF INTEGRATING AI INTO STATISTICS EDUCATION

The integration of AI into statistics education presents significant opportunities but also poses several challenges that educators and institutions must navigate (Saputra et al., 2023), (Eden et al., 2024). One major concern is access and equity. Not all students have equal access to AI-powered learning tools, and this disparity could exacerbate existing inequalities in education. Addressing this issue requires ensuring that AI tools are accessible to all learners, regardless of their socio-economic background, to promote inclusivity and fairness in education.

Another critical challenge involves data privacy and security. AI platforms often rely on the collection of student data to personalize learning experiences. This necessity raises concerns about how such data is stored and handled. It is essential for educators and developers to implement robust safeguards to protect student privacy and prevent data misuse.

Additionally, teacher training is a key factor in the effective integration of AI tools. Educators must receive adequate training and support to use these technologies effectively in their teaching practices. Without this preparation, AI tools may be underutilized or, worse, could hinder learning outcomes if implemented incorrectly. Lastly, there is the risk of over-reliance on AI in the learning process. While AI has the potential to significantly enhance statistical education, students must still develop foundational skills in statistical thinking. A balance must be struck to ensure that technology complements traditional learning methods rather than replacing them entirely.

5. FUTURE DIRECTIONS: AI-ENHANCED LEARNING FOR THE NEXT GENERATION OF STATISTICIANS

The future of statistical education is poised for a transformative evolution with the deeper integration of AI technologies, promising to deliver more personalized, interactive, and effective learning experiences. To fully harness this potential, a collaborative effort between educators and developers will be essential to create AI tools that are not only technologically advanced but also pedagogically sound, widely accessible, and ethically responsible. These tools must be designed to align with the goals of fostering critical statistical thinking while ensuring equity and inclusivity in education.

One promising direction is the development of hybrid learning environments in which AI tools augment traditional teaching methods. By combining the interactivity and adaptability of AI with the human connection and contextual understanding offered by teachers, this approach provides a balanced model of education. Students could benefit from personalized AI-driven tutorials and simulations while retaining the opportunity to engage with teachers for deeper discussions and nuanced insights. This synergy could significantly enhance both the efficiency and the quality of learning experiences.

AI also has the potential to redefine collaboration in statistical education. Future AI systems could act as facilitators in group problem-solving scenarios, encouraging students to work together to tackle complex statistical challenges. Such tools would promote not only a deeper understanding of statistical concepts but also the development of essential communication and teamwork skills. By simulating collaborative environments that mirror real-world problem-solving contexts, AI can prepare students to apply their knowledge effectively in diverse professional settings.

Moreover, the role of AI in assessments is set to become more sophisticated, enabling real-time evaluation of student progress. Advanced AI systems could design differentiated tasks based on individual performance, ensuring that students are continually challenged at an appropriate level. This adaptive assessment approach would allow for immediate feedback and tailored learning pathways, helping students address their weaknesses while building on their strengths. As AI evolves, its integration into statistical education will likely lead to a new era of learning, one that is not only more effective but also more engaging and inclusive for the next generation of statisticians.

6. CONCLUSION

AI offers exciting opportunities to redefine how we teach and learn statistics. By transforming traditional modes of representation into dynamic, interactive, and personalized experiences, AI can help students develop deeper statistical thinking and application skills. However, the successful integration of AI in statistical education requires careful consideration of challenges related to access, equity, data privacy, and teacher preparedness. As we look to the future, the collaboration between educators, technologists, and policymakers will be crucial in creating a learning environment where AI enhances—not replaces—the fundamental principles of statistical education.

REFERENCES

Alcover, R., Chirivella, V., & Vazquez, E. (2024). A proposal to speed up the learning of statistics using generative AI. *Proceedings of the 17th Annual International Conference of Education, Research and Innovation*, 3697-3701, Seville, Spain. DOI: 10.21125/iceri.2024.0944

Calonge, D. S., Smail, L., & Kamalov, F. (2023). Enough of the chit-chat: A comparative analysis of four AI chatbots for calculus and statistics. *Journal of Applied Learning & Teaching*, *6*(2). Advance online publication. DOI: 10.37074/jalt.2023.6.2.22

Eden, C.A., Chison, O.N. and Ademiyi, I.S. (2024). Integrating AI in education: Opportunities, challenges, and ethical considerations. *Magna Scientia Advanced Research and Reviews*, 10(02), 06-013. DOI: 10.30574/msarr.2024.10.2.0039

Ellis, A. R., & Slade, E. (2023). A new era of learning: Considerations for ChatGPT as a tool to enhance statistics and data science education. *Journal of Statistics and Data Science Education : An Official Journal of the of the American Statistical Association*, *31*(2), 128–133. DOI: 10.1080/26939169.2023.2223609

Gilbert, J. K. (2010). *The role of visual representations in the learning and teaching of science: An introduction. Asia-Pacific Forum on Science Learning and Teaching, 11(1).* Foreword.

Gonzalez, O., Isoda, M., & Araya, R. (2020). A new framework for statistical thinking in the time of big data and the digital economy. *Journal of Southeast Asian Education*, *1*(1), 61–70.

Hasim, S. M., Rosli, R., & Halim, L. (2024). A systematic review on teaching strategies for fostering students' statistical thinking. *International Journal of Learning. Teaching and Educational Research*, *23*(1), 136–158. Advance online publication. DOI: 10.26803/ijlter.23.1.8

Mittal, U., Sai, S., Chamola, V., & Sangwan, D. (2024). A comprehensive review on generative AI for education. *IEEE Access : Practical Innovations, Open Solutions*, *12*, 142733–142759. DOI: 10.1109/ACCESS.2024.3468368

Morales-Vargas, S. R., & Benitez-Medina, R. (2024). The use of artificial intelligence as a learning tool. A first focus on solving statistics problems. *Proceedings of the 17th Annual International Conference of Education Research and Innovation*, 3412-3416, Seville, Spain. DOI: 10.21125/iceri.2024.0882

Quinones, D., Ruz, F., Diaz-Arancibia, J., Paz, F., Osega, J., & Rojas, L. F. (2024). Innovating statistics education: The design of a novel app using design thinking. *Applied Sciences (Basel, Switzerland), 14*(18), 8515. DOI: 10.3390/app14188515

Rane, N. L., Choudhary, S. P., & Rane, S. (2024). Education 4.0 and 5.0: Integrating artificial intelligence (AI) for personalized and adaptive learning. *Journal of Artificial Intelligence and Robotics, 1*(1), 29–43. DOI: 10.61577/jaiar.2024.100006

Saputra, I., Astuti, M., Sayuti, M., & Kusumastuti, D. (2023). Integration of artificial intelligence in education: Opportunities, challenges, threats and obstacles. A literature review. *The Indonesian Journal of Computer Science, 12*(4), 1590–1600. DOI: 10.33022/ijcs.v12i4.3266

Shaik, T., Tao, X., Li, Y., Dann, C., McDonald, J., Redmond, P., & Galligan, L. (2022). A review of the trends and challenges in adopting natural language processing methods for education feedback analysis. *IEEE Access : Practical Innovations, Open Solutions, 10*, 56720–56739. DOI: 10.1109/ACCESS.2022.3177752

Souravlas, S., Anastasiadou, S., Economides, Th., & Katsavounis, S. (2023). Probabilistic community detection in social networks. *IEEE Access : Practical Innovations, Open Solutions, 11*, 25629–25641. DOI: 10.1109/ACCESS.2023.3257021

Souravlas, S., Anastasiadou, S., & Katsavounis, S. (2021). More on pipelined dynamic scheduling of big data streams. *Applied Sciences (Basel, Switzerland), 11*(1), 61. DOI: 10.3390/app11010061

Zhai, X., Chu, X., Chai, C. S., Jong, M. S. Y., Istemic, A., Spector, M., Liu, J. B., & Li, J. Y. Y. (2021). A review of artificial intelligence (AI) in education from 2010 to 2020. *Complexity, 1*(1), 8812542. DOI: 10.1155/2021/8812542

Chapter 3
Principal Component Analysis vs. Implicative Statistical Analysis:
The Case of Probability Distributions' Understanding

Sofia Anastasiadou
https://orcid.org/0000-0001-6404-5003
University of Western Macedonia, Greece

ABSTRACT

The present study explores the similarities and differences between two distinct statistical methods: Principal Component Analysis (PCA) and Implicative Statistical Analysis. Specifically, the study assesses the results derived from applying these two methodologies to a case study on Greek students' understanding of probability distributions (e.g., Binomial Distribution, Poisson Distribution, Normal Distribution). The research involved 119 second-year students from the Department of Educational and Social Policy at the University of Macedonia. The primary objective was to determine whether these two statistical approaches yield similar or different conclusions regarding student comprehension. Additionally, the specific advantages of each statistical method are discussed in detail.

INTRODUCTION

Learning probability at the high-school or introductory college level can be challenging, as students often hold intuitive yet incorrect perceptions of chance, which conflict with formal principles of probability (Anastasiadou, 2009a; Anastasiadou &

DOI: 10.4018/979-8-3693-9934-7.ch003

Loukas, 2009). These misconceptions may be influenced by negative attitudes toward probability and statistics (Anastasiadou, 2005a; 2005b; 2005c; 2007). Additionally, attitudes of both students and teachers regarding the use of various representational methods in teaching and learning statistics are particularly important, as they can contribute significantly to difficulties in understanding (Anastasiadou, 2019a; 2019b).

In this section, we identify common errors and misconceptions in fundamental areas of probability, explain the underlying causes of these misunderstandings, and provide illustrative examples. Recognizing these recurring difficulties enables educators to design more effective instructional strategies to proactively address student challenges and enhance comprehension (Chadjipantelis & Anastasiadou, 2010). Furthermore, highlighting these persistent issues empowers educators to develop targeted teaching methods, effectively addressing typical student pitfalls and significantly improving their understanding of and confidence in probability concepts (Loukas et al., 2008; Anastasiadou, 2002, 2013a, 2024; Anastasiadou et al., 2013b).

Even fundamental probability concepts, such as determining when to add or multiply probabilities, are commonly misapplied. Frequent mistakes include:

Adding probabilities when "and" is intended: Students sometimes incorrectly add probabilities for events that should instead be multiplied ("and" scenarios).

Not subtracting overlaps in "or" problems: Students often make the error of directly adding probabilities, $P(A) + P(B)$, when calculating the probability of one event or another ("A or B"), neglecting to subtract their intersection.

Conditional probability ($P(A|B)$) also frequently causes confusion. Students typically struggle to interpret conditional statements correctly and often misapply the definition of conditional probability.

Regarding Binomial distribution misconceptions, several common issues arise. Firstly, students often misidentify the number of trials, causing confusion over the total number of trials or mistakenly assuming an infinite number of trials. Additionally, students may incorrectly apply Binomial distribution in situations where the probability of success is not consistent across trials, reflecting a misunderstanding of the assumption of equal probability. Another significant misconception involves independence of trials, where students erroneously treat dependent events as independent, leading to incorrect use of the Binomial formula.

Regarding Poisson distribution misconceptions, students frequently encounter several difficulties. Firstly, students mistakenly apply the Poisson distribution to events that are not independent or have non-constant rates, violating key assumptions of the distribution. Furthermore, students often confuse discrete with continuous distributions, incorrectly applying Poisson to continuous data sets. Another common error involves the incorrect identification of relevant intervals or time periods, reflecting difficulty in properly defining the conditions required for a Poisson distribution.

Regarding Normal distribution misconceptions, students frequently encounter several difficulties. Students often assume that all datasets are normally distributed without verifying this assumption through tests or graphical analysis. Additionally, they may misinterpret and incorrectly apply the Empirical Rule (68-95-99.7 rule) to datasets that are not symmetric or normally distributed. Furthermore, students commonly confuse sample means and standard deviations with the actual parameters of the Normal distribution, indicating confusion between population parameters and sample statistics.

Furthermore, students commonly confuse sample means and standard deviations with the actual parameters of the Normal distribution, indicating confusion between population parameters and sample statistics. Another common misconception includes misunderstanding the concept of standardization (Z-scores), resulting in miscalculations and incorrect interpretations of probabilities associated with specific outcomes. Additionally, students may fail to distinguish clearly between the Normal distribution and other continuous distributions, applying Normal distribution methods incorrectly to data that do not follow a Normal pattern.

A central question in statistics is determining which statistical methodology is most suitable for each specific case. Another important consideration is whether different statistical methods used in data analysis lead to consistent results and thus to similar conclusions. Therefore, the current study aims to evaluate two distinct statistical methods—Principal Component Analysis (PCA) and Implicative Statistical Analysis (ISA)—by comparing their outputs in a case study that investigates Greek students' understanding of probability distributions (Fotiadis & Anastasiadou, 2018a; 2018b). The study highlights both the differences and similarities between these two methodologies and provides a detailed discussion of the advantages associated with each statistical method.

2. METHODOLOGY

2.1. Statistical methodology

Principal Component Analysis (PCA) with Varimax Rotation, which produces dimensions of differentiation, was employed to examine the construct validity of the test (Anastasiadou, 2009b, 2018). Two statistical criteria were applied to determine the suitability of the sub-scales for factor analysis (Anastasiadou et al., 2010a; 2010b). The first criterion, Bartlett's Test of Sphericity, examines whether the subscales of the measurement tool are interdependent. The second criterion,

Kaiser-Meyer-Olkin (KMO) Measure of Sampling Adequacy (Kaiser, 1974), assesses the adequacy of the sample.

The main method for extracting factors was Principal Component Analysis combined with Varimax orthogonal rotation (Right-angled Rotation of Maximum Variance). Varimax rotation maximizes the variance of factor loadings, resulting in large loadings becoming larger, small loadings smaller, and intermediate loadings minimized (Hair et al., 2005). This ensures that the extracted factors (components) are linearly uncorrelated (Anastasiadou, 2006). The criterion for determining the number of retained factors was based on eigenvalues greater than 1 (Kaiser, 1960; Sharma, 1996; Hair et al., 1995).

The acceptance of the factor model relied on two conditions: each variable (item) had to load significantly (a) greater than 0.5 on one factor and (b) less than 0.4 on all other factors (Schene et al., 1998). Furthermore, each factor was required to consist of at least three variables. Additionally, variables with higher communalities (h^2) indicated a greater contribution to the factorial structure (Hair et al., 2005).

Data analysis and verification of the test's factorial structure were performed using SPSS software, version 25.

2.2. Implicative Statistical Analysis

Implicative Statistical Analysis (SIA) is a method of data analysis dedicated to identifying and structuring quasi-implicative relationships, originally developed by Gras (Gras & Kountz, 2008; Gras & Bodin, 2017; Gras et al., 1997). According to Coutourier (2008), ISA establishes the following properties among the variables it examines:

1. Relationships among variables that are asymmetrical.
2. Association measures that are non-linear and probability-based.
3. Graphical representations that follow the semantic logic of these relationships.

For analyzing the data collected in this study, Hierarchical Clustering of Variables and Gras's Implicative Statistical Analysis were conducted using specialized software known as C.H.I.C. (Classification Hiérarchique, Implicative et Cohésitive) (Bodin, Coutourier & Gras, 2000). Specifically, Similarity diagrams, Hierarchical diagrams, and Implicative diagrams were produced through applying the C.H.I.C. software to the research data. C.H.I.C. facilitates the extraction of association rules from given datasets. Utilizing implication intensity and similarity intensity, the software generates two types of hierarchical trees and an implication graph (Coutourier & Gras, 2005; Coutourier, 2008).

The first type, the similarity tree, relies on the similarity index defined by Lerman (1981) and provides a non-oriented hierarchical classification. Similarly, implication intensity is utilized to construct an oriented hierarchical tree, as well as an implication graph. These analytical methods identify similarity connections and implicative relationships among variables. Specifically, hierarchical clustering of variables is a classification procedure aimed at finding increasingly broader partitions of a variable set V, arranged in ascending order. These nested partitions are visually represented through hierarchical tree diagrams that reflect statistical similarity criteria among the variables. Similarity is determined by cross-comparing a group of variables (V) with a group of individuals or objects (E). This form of analysis allows researchers to interpret data in terms of typology and decreasing levels of similarity. Variable clusters created at specific levels in the diagram can be compared to each other (Gras & Kuntz, 2008; Gras & Bodin, 2017).

The hierarchical similarity tree is built using the following process: initially, the two most similar variables based on the similarity index are grouped together at the highest similarity level. This initial grouping is then linked either to another individual variable or to another variable group at progressively lower similarity levels, continuing until the similarity or cohesion between variables or groups becomes negligible. Shorter vertical lines in these diagrams reflect stronger connections.

The implicative diagram, resulting from applying Gras's statistical implicative method, captures directional relationships, demonstrating whether success in performing one task implies success in another related task (Anastasiadou, 2010, 2019c; Anastasiadou et al., 2024). Similarly, the implicative graph illustrates these directional implications, indicating whether successful completion of one task leads to success in another associated task (Ntotsi & Anastasiadou, 2019a; 2019b).

2.3. Sample

The study sample consisted of 119 second-year students enrolled in the Department of Educational and Social Policy at the University of Macedonia. These students attended a compulsory course covering topics in Discrete Mathematics, Probability Properties, Probability Distributions (e.g., Binomial, Poisson, Normal distributions), and Inferential Statistics. During the first semester, the students had completed and been examined in Descriptive Statistics, which was also compulsory. Following instruction on Probability Distributions, all 119 students participated in a term examination.

2.4. Examination Task-Exercises

The task consisted of nine exercises related to probability distributions (Binomial, Normal, and Poisson distributions). Each exercise (variable) was coded with an uppercase 'V' (Variable), followed by a number indicating the exercise, and lowercase letters denoting sub-questions within the exercise. Specifically, the first lowercase letter represents the first sub-question, and the second lowercase letter represents the second sub-question. Student responses were scored as follows: 1 for a fully correct answer, 0.5 for a partially correct answer, and 0 for incorrect or unanswered questions.

Exercise 1 (V1nn):

"We assume that a family's monthly expenses for children's activities (e.g., ballet, sports, foreign languages, etc.) and children's personal expenses (pocket money) follow a normal distribution with a mean of €121 and a standard deviation of €9. Estimate the probability that:

i) the family's monthly expenses are less than €121, and
ii) the family's monthly expenses range between €112 and €130.

Exercise 2 (V2nn): The weight of six-year-old girls follows a normal distribution with a mean of 25 kg and a standard deviation of 5 kg.

i) What is the probability that a randomly chosen girl's weight does not exceed 25 kg?
ii) What is the probability that a girl's weight is between 20 kg and 30 kg?

Exercise 3 (V3nn): The height of six-year-old boys follows a normal distribution with a mean of 125 cm and a standard deviation of 3 cm.

i) What is the probability that a randomly chosen boy's height does not exceed 119 cm?
ii) What is the probability that a boy's height is between 125 cm and 134 cm?

Exercise 4 (V4bb): Approximately 60% of five-year-old preschool children watch and enjoy the cartoon "Pink Panther." Suppose we randomly ask 4 preschool children about their favorite TV series.

i) What is the probability that none of these preschool children watch "Pink Panther"?
ii) What is the probability that exactly one child watches "Pink Panther"?

Exercise 5 (V5bb): About 40% of six-year-old preschool children can write their names. We randomly select 4 preschool children.

i) What is the probability that none of the selected children know how to write their name?
ii) What is the probability that exactly one child knows how to write his or her name?

Exercise 6 (V6nb): The weight of six-year-old boys follows a normal distribution with a mean of 40 kg and a standard deviation of 9 kg.

i) What is the probability that a randomly selected boy's weight does not exceed 49 kg?
ii) Suppose we randomly select 4 boys. What is the probability that exactly one of these boys has a weight not exceeding 49 kg?

Exercise 7 (V7pp): A school director receives, on average, 4 calls per hour from parents. Calculate the following probabilities:

i) That no calls are received in the next hour.
ii) That more than 2 calls are received in one hour.

Exercise 8 (V8pp): Historically, an average of 2 students fail their mathematics final examination each year. Assume that the number of failures follows a Poisson distribution with $\lambda = 2$.

i) What is the probability that no students fail this year's final exam?
ii) What is the probability that at least one student fails this year's final exam?

Exercise 9 (V9pp): A teacher receives, on average, 6 parent visits per hour. Calculate the probabilities:

i) That no visits are received in a twenty-minute interval.
ii) That more than 4 visits are received in a half-hour interval.

3. RESULTS

3.1. Principal Components Analysis Results

The sampling adequacy index (KMO = 0.739 > 0.60) indicated that the data collected were suitable for factor analysis. Bartlett's test of sphericity ($\chi^2 = 499.723$, df = 10, $p < 0.001$) further confirmed that principal component analysis was appropriate. The total variance explained by the extracted factors exceeded 70%, specifically 74.905%. Using the eigenvalue criterion (eigenvalues greater than 1), two factors were extracted from the data analysis, accounting collectively for 74.905% of the total variance. After Varimax rotation (Rotation Sums of Squared Loadings), the first factor (F1) explained 45.631% of the total variance, while the second factor (F2) accounted for an additional 29.272%. Moreover, the eigenvalues of these factors (3.389 for F1 and 3.873 for F2) exceeded 1, confirming the satisfactory interpretability and robustness of the factor structure.

Table 1 provides information regarding the communalities, which indicate the quality of the variables. Specifically, when communalities (joint factor variance) for each item exceed 0.50, this demonstrates satisfactory measurement quality. In particular, the communalities for variables V1nn, V3nn, V2nn, V5bb, V4bb, V6nb, V7pp, V8pp, and V9pp are 0.821, 0.784, 0.596, 0.678, 0.869, 0.863, 0.838, 0.567, and 0.620 respectively.

Additionally, all variable loadings are greater than 0.40, further confirming the reliability and quality of the test items. Each variable exhibits stronger correlations with variables within the same factor compared to those belonging to different factors, loading primarily on a single factor axis. Given that most communalities exceed the threshold of 0.50, the measurement quality is confirmed to be satisfactory. Consequently, the two-factor model derived from the sample exhibits strong factorial validity.

Table 1. Principal components analysis results

	Factors		
Items	**F1**	**F2**	**Communality**
V1nn	0.818		0.821
V3nn	0.807		0.784
V2nn	0.797		0.596
V5bb	0.781		0.678
V4bb	0.758		0.869

continued on following page

Table 1. Continued

	Factors		
Items	**F1**	**F2**	**Communality**
V6nb	0.721		0.863
V7pp		0.796	0.838
V8pp		0.765	0.567
V9pp		0.709	0.620
Eigenvalue	3.389	3.873	
Variance Explained %	45.631%	29.272%	
Total Variance Explained %	74.903%		

Kaiser-Meyer-Olkin Measure of Sampling Adequacy = 0.739
Bartlett's Test of Sphericity: x^2=499.723, p=0.000

More specifically, based on students' responses and factor analysis results, the first factor (F1), which accounts for 45.631% of the total variance after Varimax rotation, includes variables V1nn, V3nn, V2nn, V5bb, V4bb, and V6nb. These variables exhibit high factor loadings of 0.818, 0.807, 0.797, 0.781, 0.758, and 0.720 respectively. Variables associated with Normal and Binomial distributions are grouped within Factor F1, clearly indicating that students approach Normal and Binomial distribution problems in a similar, distinct manner.

The second factor (F2), accounting for 29.272% of total variance after Varimax rotation, includes variables V7pp, V8pp, and V9pp, with high loadings of 0.796, 0.765, and 0.709 respectively. Variables related to the Poisson distribution are distinctly grouped within Factor F2.

The strong correlation among variables V7pp, V8pp, and V9pp within their respective factors, as well as their distinct separation from Factor F1, confirms that the two factorial axes are independent of each other. This indicates that students apply similar strategies when solving problems involving Normal and Binomial distributions, whereas they utilize completely different approaches for problems involving the Poisson distribution.

3.2. Implicative Statistical Analysis Results

Observations on the Similarity Diagram: In the similarity diagram (Figure 1), the grouping of items is based on students' responses to the nine exercises included in the assessment. More specifically, the grouping refers to the items measuring students' cognitive abilities related to solving exercises involving probability distributions.

The identified similarities among these variables are statistically significant at a 99% confidence level. From the diagram, two distinct similarity subgroups (Subgroup A and Subgroup B) can be observed.

Figure 1. Similarity diagram

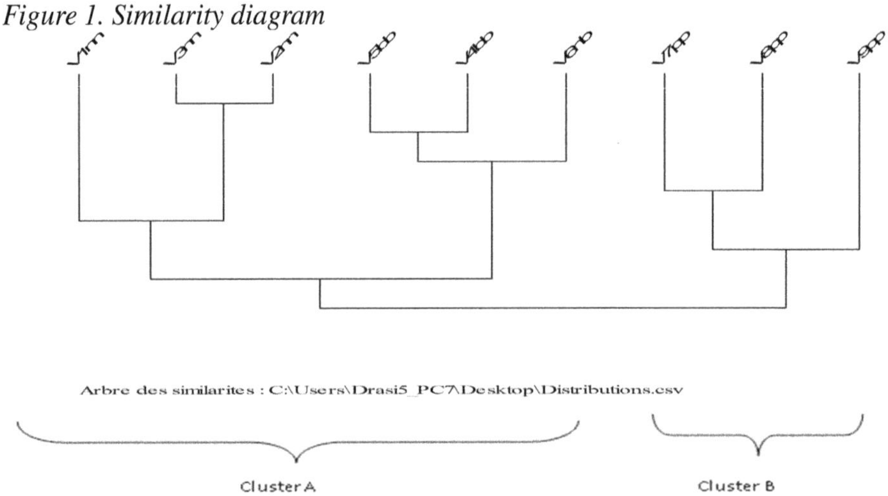

The first subgroup (A) refers to similarity relationships among variables (V1nn, V3nn, V2nn, V5bb, V4bb, V6nb) associated with problems involving Normal and Binomial Distributions. The similarity observed within this subgroup indicates that students approach the statements of variables V1nn, V3nn, V2nn, V5bb, V4bb, and V6nb in a similar manner.

Specifically, a notably strong similarity is evident between items V3nn and V2nn (Classification au niveau 1: (V3nn V2nn); similarity: 0.967933). Moreover, the similarity among variables V1nn, V3nn, and V2nn as a subgroup is also significant (Classification au niveau 5: (V1nn (V3nn V2nn)); similarity: 0.728706).

Additionally, the similarity between variables V5bb and V4bb is strong (Classification au niveau 2: (V5bb V4bb); similarity: 0.935091). The broader subgroup, including these two variables along with V6nb, also exhibits a marked similarity (Classification au niveau 3: ((V5bb V4bb) V6nb); similarity: 0.874396).

Overall, the collective similarity among all variables V1nn, V3nn, V2nn, V5bb, V4bb, and V6nb is moderately strong (Classification au niveau 7: ((V1nn (V3nn V2nn)) ((V5bb V4bb) V6nb)); similarity: 0.583577).

The second subgroup, B, refers to similarity relations among the variables V7pp, V8pp, and V9pp, which address problems related to the Poisson Distribution. This similarity among variables in subgroup B indicates that students handle the statements of V7pp, V8pp, and V9pp similarly.

Specifically, within this subgroup, there is a strong similarity between items V7pp and V8pp (Classification at level 4: (V7pp V8pp) similarity: 0.773732). The similarity across all items of this subgroup is also notable (Classification at level 6: ((V7pp V8pp) V9pp) similarity: 0.628865).

In contrast, the similarity between subgroup A and subgroup B is extremely weak (Classification at level 8: (((V1nn (V3nn V2nn)) ((V5bb V4bb) V6nb)) ((V7pp V8pp) V9pp)) similarity: 0.0311883). This very weak similarity indicates that subgroup A and subgroup B are distinctly separate clusters. Students approach Normal and Binomial Distribution problems similarly but handle Poisson Distribution problems in a distinctly different manner. This finding aligns with results obtained through the Principal Components Analysis (PCA) application.

Observations on the Hierarchical Diagram: In the similarity diagram (Figure 2), hierarchical relationships among variables are presented clearly, in order of importance. Furthermore, this hierarchical representation allows identification of the direction of these relationships. The observed implications are statistically significant at the 99% confidence level.

Figure 2. Hierarchical diagram

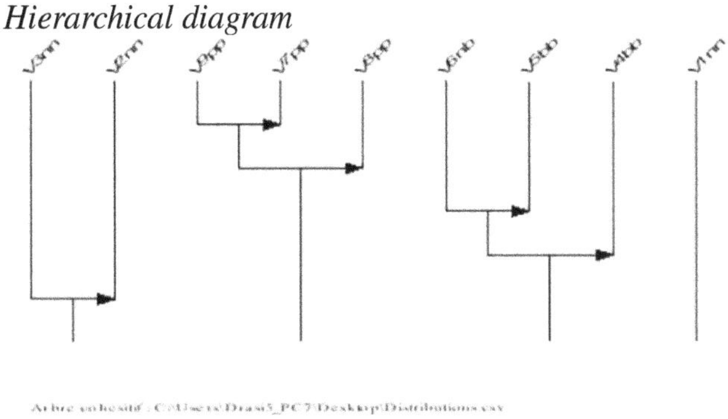

The first hierarchical group consists of items V3nn and V2nn (cohesion: 0.636). Specifically, in the hierarchy V3nn–V2nn (cohesion: 0.636), the response to item V3nn implies the response to item V2nn. It is important to highlight the significance of this hierarchical relationship.

The second hierarchical group includes items V9pp–V7pp–V8pp (cohesion: 0.988), indicating that the response to item V9pp leads directly to the responses for items V7pp and V8pp. Within this group, the partial hierarchy between V9pp and V7pp (cohesion: 0.996) is notably strong.

Another hierarchical group involves items V6nb–V5bb–V4bb (cohesion: 0.785). This group demonstrates that the response to item V6nb guides the response to V5bb, and both subsequently influence the response to V4bb. Particularly noteworthy is the hierarchy between V6nb and V5bb (cohesion: 0.937).

It is also important to note that item V1nn does not exhibit hierarchical relationships with any other items. The cohesions described above provide insights into students' cognitive skills and their capabilities in solving probability problems.

4. CONCLUSIONS / DISCUSSION

In the present paper the advantages of two distinct statistical methodologies are discussed in detail. More specifically, PCA and Implicative Statistical Analysis were put into action. The implementation of both methods demonstrated the way students' responses make calculations regarding Probability Distribution problems. Comparing the two, PCA and Implicative Statistical Analysis we realize that we had the same grouping of the variables connected with Normal, Binomial and Poisson Distribution' problems. In addition, following the application of PCA the factorial structure of items was determined and the significant variables of each Probability category occurred and lead to components formulation. Through Implicative Statistical Analysis, the manner, consistency and cohesion with which the students respond to the variables, were designated and consequently, their cognitive competences was exposed. Additionally, the application of Implicative Statistical Analysis brought out the hierarchy among the items and therefore their difficulty. In conclusion, the application of the two distinct methods allows data analysis and describes different aspects of the results. Subsequently, their combination offers a more detailed and complete analysis and scientific conclusions. Students seem to adopt specific problem-solving methods for each type of probability distribution but often encounter difficulties transitioning effectively between different distributions. Addressing these misconceptions about probability distributions requires further research, particularly through the integration of artificial intelligence (Souravlas & Anastasiadou, 2022; Souravlas et al., 2020), cloud computing (Souravlas et al., 2021a; 2021b; 2021c), advanced technologies, and innovative educational tools. Such research is essential, considering the importance and inevitability of thoroughly understanding probability distributions (Souravlas et al., 2022a; 2022b; 2023).

REFERENCES

Anastasiadou, S. (2002). Enhancing prospective teachers' knowledge of students' attitudes and representations towards statistics-the case of Greek students. In the Journal 'Scientia Paedacogika Experimentals'. *XXXIX*, *1*, 103–127.

Anastasiadou, S. (2004). Teacher's opinions about statistics and probability. In the journal 'Quaderni di Ricerca in Didattica', no 14. pp. 134-143. http://math.unipa .it/~grim/quaderno14.htm

Anastasiadou, S. (2005a). Survey of attitudes and beliefs toward Statistics in Greek Secondary Education International Conference on Mathematics Education, Svishtov, Bulgaria, pp. 125-133.

Anastasiadou, S. (2005b). Affective reactions and attitudes of the last class of Greek high school students towards statistics, In M Boosch (Eds) Proceedings 0r the Fourth Conference of the European Society for Research in Mathematics Education (CERME 4), Sant Feliu de Guíxols, Spain, 2005 in the http://cerme4.crm.es/Papers %20definitius/5/Anastasiadou%20Sofia.pdf

Anastasiadou, S. (2005c). Survey of attitudes and beliefs toward Statistics in Greek Secondary Education International Conference on Mathematics Education, Svishtov, Bulgaria, pp. 125-133.

Anastasiadou, S. (2007). Affective Responses and Attitudes of ninth grade Greek Students. *Proceedings of the 5th Mediterranean Conference in Mathematics Education*, Rodes, Greece pp. 169-178.

Anastasiadou, S. (2009a). Greek students' ability in probability problem solving. In the Sixth Conference of the European Society for Research in Mathematics Education. Proceedings of CERME 6, Lyon France © INRP 2010, pp. 404-412. https://ife.ens-lyon.fr/editions/editions-electroniques/cerme6/working-group-3

Anastasiadou, S. (2009b). Detection of Factors Influencing Greek Technological Institute Students' Choice of Study Department with the Aid of Correspondence Factor Analysis and Path Analysis. *Journal of the World Universities Forum*, *2*(4), 131–142. DOI: 10.18848/1835-2030/CGP/v02i04/56606

Anastasiadou, S. (2010). Pre-service teachers' performance on the learning of probability distributions and the role of projects: A multilevel statistical analysis. In Regnier, Jean-Claude, Spagnolo, Filippo, Di Paola, Benedetto, & Gras, Regis, (Eds). *5th International Conference Implicative Statistic Analysis*, Palermo, Italy pp 73-84.

Anastasiadou, S. (2018). Comparison of multivariate methods in group/cluster identification. Dissertation thesis of the degree of MSc. in Research Methodology in Biomedicine, Biostatistics and Clinical Bioinformatics, Fuculty of Medicine, University of Thessaly.

Anastasiadou, S. (2019a). Pre-service teachers' attitudes toward the use of different modes of representation in statistics teaching and learning. 11th annual International Conference on Education and New Learning Technologies Palma de Mallorca, Spain, EDULEARN 19. pp. 1-5.

Anastasiadou, S. (2019b). Pre-service teachers' attitudes toward the use of different modes of representation in statistics teaching and learning. 11th annual International Conference on Education and New Learning Technologies Palma de Mallorca, Spain, EDULEARN 19. pp. 1-5.

Anastasiadou, S. (2019c). Comparison of contemporary advanced statistical methods regarding construct validity evaluation of TEIque-SF instrument: Statistical Implicative Analysis vs. Principal Components Analysis. *9ème Colloque International sur Analyse Statistique Implicative (ASI 10).* Belfort – France. pp. 148-163.

Anastasiadou, S., Anastasiadis, L., Angeletos, T., & Vandikas, J. (2010b). A Multidimensional Statistical Analysis of Students' Attitudes toward Physics. *The International Journal of Diversity in Organisations, Communities and Nations, 16*(5), 341–356. DOI: 10.18848/1447-9532/CGP/v10i01/39794

Anastasiadou, S., Anastasiadis, L., Vandikas, J., & Angeletos, T. (2010a). Implicative statistical analysis and Principal Components Analysis in recording students' attitudes toward electronics and electrical constructions subject. *International Journal of Technology, Knowledge and Society, 16*(5), 341–356.

Anastasiadou, S., Chadjipadelis, T., & Kofou, I. (2013a). The role of projects in helping Greek university students from Pre-school Education department to understand probability distributions International Symposium Elementary Mathematics Teaching (SEMT '13). Prague, Czech Repuplic. International Symposium Elementary Maths Teaching. Proceedings: Tasks and tools in elementary mathematics, Prague [2013] σελ.67-74.

Anastasiadou, S., Chadjipadelis, T., & Kofou, I. (2013a). Pre-school Education department to understand probability distributions. International Symposium Elementary Mathematics Teaching (SEMT '13), Prague, Czech Republic, In Jarmila Novotná and Hana Moraová (Eds). pp 67-75.

Anastasiadou, S., & Loukas, D. (2009). Greek pre-service teachers' cognitive abilities in understanding the concept of frequency: A multilevel statistical analysis. *International Journal of Learning*, *16*(5), 189–202. DOI: 10.18848/1447-9494/CGP/v16i05/46268

Anastasiadou, S., Zirinoglou, P., Seremeti, L., Masouras, A., Komodromos, M. & Anastasiadis, L. (2024). Statistical Implicative Analysis: A Distinguished Statistical Methodology in Evaluating Greek Students' Attitudes Toward the Use of AI Source Title: Real-World Tools and Scenarios for Entrepreneurship Exploration, (2024), 79-100, .DOI: 10.4018/979-8-3693-3100-2.ch006

Bodin, A., Courtourier, R., & Gras, R. (2000). *CHIC: Classification Hiérarchique Implicative et Cohesitive-Version sous Windows – CHIC 1.2*. Association pour le Recherche en Didactique des Mathématiques.

Chadjipantelis, T., & Anastasiadou, S. (2010). Pre-service teachers' understanding of probability Distributions: A Multilevel Statistical Analysis. International Conference on the Teaching of Statistics. ICOTS 8, Ljubljanja, Slovenia 2010. https://www.stat.auckland.ac.nz/~iase/publications.php?show=icots8

Coakes, S., & Steed, L. (1999). *SPSS Analysis without Anguish*. John Willey & Sons.

Courturier, R. (2008). CHIC: Cohesive Hierarchical Implicative Classification. In Gras, R., Suzuki, E., Guillet, F., & Spagnolo, F. (Eds.), *Statistical Implicative Analysis* (pp. 1–13). Springer Verlag Berlin Heidelberg. DOI: 10.1007/978-3-540-78983-3_2

Coutourier, Raphael., & Gras, Rgis. (2005). CHIC: Traitement de données avec L; analyse implicative Extraction et Gestion des Connaissances, Volume II, RMTI, Cepadues, Paris, pp. 679-684.

Fotiadis, Th., & Anastasiadou, S. (2018a). Contemporary advanced statistical methods for the science of marketing: Principal Components Analysis vs Analysee Factorielle des Correspondances. Proceeding of the ICME 2018- 1st International Conference on Marketing and Entrepreneuship. Cyprus. https://hephaestus.nup.ac.cy/handle/11728/11392

Fotiadis, Th., & Anastasiadou, S. (2018b). Contemporary advanced statistical methods for the science of marketing: Implicative Statistical Analysis vs Principal Components Analysis. Proceeding of the ICME 2018- 1st International Conference on Marketing and Entrepreneuship. Cyprus. https://hephaestus.nup.ac.cy/handle/11728/11393

Gras, R., & Bodin, A. (2017). L'A.S.I., Analyseur et révélateur de la complexité cognitive taxonomique. 9ème Colloque International sur Analyse Statistique Implicative, Belfort – France, In Jean-Claude Régnier, Régis Gras, Raphaël Coutourier, Antoine Bodin (Eds.), pp. 128-142.

Gras, R., & Kuntz, P. (2008). An overview of the Statistical Implicative Analysis (SIA) development. In Gras, R., Suzuki, E., Guillet, F., & Spagnolo, F. (Eds.), *Statistical Implicative Analysis* (pp. 11–40). Springer Verlag Berlin Heidelberg. DOI: 10.1007/978-3-540-78983-3_1

Gras, R., Peter, P., Briand, H., & Philippé, J. (1997). Implicative Statistical Analysis. In C. Hayashi, N. Ohsumi, N. Yajima, Y. Tanaka, H. Bock, Y. Baba (Eds.). Proceedingsofthe5th Conference of the International Federation of Classication Societies, Volume 2, pp.412-419. Tokyo, Berlin, Heidelberg, New York: Springer-Verlag.

Hair, J., Anderson, R., Tatham, R., & Black, W. (1995). *Multivariate Data Analysis With Readings*. Prentice-Hall International, Inc.

Hair, J., Black, B., Babin, B., Anderson, E. R., & Tatham, R. L. (2005). *Multivariate Data Analysis*. Pearson Education Inc.

Kaiser, H. F. (1960). The application of electronic computers to factors analysis. *Educational and Psychological Measurement*, *20*(1), 141–151. DOI: 10.1177/001316446002000116

Kaiser, H. F. (1974). An index of vactorial simplicity. *Psychometrika*, *39*(1), 31–36. DOI: 10.1007/BF02291575

Lerman, I. C. (1981). *Classification et analyse ordinale des donnes*. Dunod.

Loukas, D., Papadimitriou, I., Drosos, G., & Anastasiadou, S. (2008). Investigation of a categorical data distribution via the methods of multivariate data analysis – An application to a stochastic multicriteria group decision-making problem. 67th Meeting of the European Working Group "Multiple Criteria Decision Aiding", Rovaniemi, Lapland, Finland.

Ntotsi, P., & Anastasiadou, S. D. (2019a). Contemporary advanced statistical methods for the science of educational research: Principal components analysis versus L' analysee factorielle des correspondances. *New Trends and Issues Proceedings on Humanities and Social Sciences*. [Online]. 6(1), pp 246–255. Available from from: www.prosoc.eu

Ntotsi, P., & Anastasiadou, S. D. (2019b). Comparison of multivariate pattering methods in group/cluster identification regarding the science of educational research: Implicative Statistical Analysis vs. L' Analysee Factorielle des Correspondances. *New Trends and Issues Proceedings on Humanities and Social Sciences.* [Online]. 6(1), pp 238–245. www.prosoc.eu

Schene, A., Wijngaarden, B., & Koeter, M. (1998). Family Care giving in Schizophrenia: Domains, Distress. *Schizophrenia Bulletin, 24*(4), 609–618. DOI: 10.1093/oxfordjournals.schbul.a033352 PMID: 9853792

Sharma, S. (1996). *Applied Multivariate Techniques.* John Willey & Sons, Inc.

Souravlas, S., & Anastasiadou, S. (2020). Pipelined Dynamic Scheduling of Big Data Streams. *Applied Sciences (Basel, Switzerland), 10*(14), 4796. DOI: 10.3390/app10144796

Souravlas, S., & Anastasiadou, S. (2022). On Implementing Social Community Clouds Based on Markov Models. *IEEE Transactions on Computational Social Systems, 2022.* Advance online publication. DOI: 10.1109/TCSS.2022.3213273

Souravlas, S., Anastasiadou, S., Economides, Th., & Katsavounis, S. (2023). Probabilistic Community Detection in Social Networks. IEEE Access, Journal: IEEE Access, Publication Date: 2023, Volume: 11, On Page(s): 25629-25641, Print ISSN: 2169-3536, Online ISSN: 2169-3536, Digital Object Identifier: DOI: 10.1109/ACCESS.2023.3257021

Souravlas, S., Anastasiadou, S., & Katsavounis, S. (2021b). A Survey on the Recent Advances of Deep Community Detection. *Applied Sciences (Basel, Switzerland), 11*(16), 7179. DOI: 10.3390/app11167179

Souravlas, S., Anastasiadou, S., & Katsavounis, S. (2021c). More on Pipelined Dynamic Scheduling of Big Data Streams. *Applied Sciences (Basel, Switzerland), 11*(1), 61. DOI: 10.3390/app11010061

Souravlas, S. Anastasiadou, S. & Kostoglou, I. (2022b). A Novel Method For General Hierarchical System Modeling Via Colored Petri Nets Based on Transition Extractions From Real Data Sets. Appl. Sci. 2023, 13, 339. DOI: 10.3390/app13010339

Souravlas, S., Anastasiadou, S., Tantalaki, N., & Katsavounis, S. (2021a). A Dynamic Load Balanced Task Distribution Strategy for Heterogeneous Cloud Platforms Based on Markov Process Modeling. *IEEE Access : Practical Innovations, Open Solutions, 10*, 2169–3536. https://ieeexplore.ieee.org/abstract/document/9729723

Souravlas, S., Anastasiadou, S. D., Tantalaki, N., & Katsavounis, S. (2022a). A Fair, Dynamic Load Balanced Task Distribution Strategy for Heterogeneous Cloud Platforms Based on Markov Process Modeling," in IEEE Access, vol. 10, pp. 26149-26162, 2022, DOI: 10.1109/ACCESS.2022.3157435

Souravlas, S., Katsavounis, S., & Anastasiadou, S. (2020). On Modeling and Simulation of Resource Allocation Policies in Cloud Computing Using Colored Petri Nets. *Applied Sciences (Basel, Switzerland)*, *10*(16), 5644. DOI: 10.3390/app10165644

Chapter 4
Python-Powered Set Operation:
An Interactive Web Tool for Discreate Mathematics Education and Application

Atharva Saraf
https://orcid.org/0009-0006-3842-5508
Ajeenkya D.Y. Patil University, India

Susanta Das
https://orcid.org/0000-0002-9314-3988
Ajeenkya D.Y. Patil University, India

ABSTRACT

This chapter shows the development, implementation, & evaluation of comprehensive desktop-based software tool engineering using Python to facilitate an array of set operations. Software furnishes users with a versatile platform for conducting fundamental set manipulations (e.g., union, intersection, computation, difference, symmetric different, power set generation, & Cartesian product computation). It features a user-friendly interface with input fields for two sets with clearly labelled buttons for each operation. This design aims to make the software easy to use for everyone, from beginners to experts, enhancing accessibility and user satisfaction. Through judicious utilization of python functions skillfully designed to capture the complexities of each set operation, it ensures seamless execution and expedient delivery of precise results in real-time. Algorithms supporting each operation are thoroughly explained, clarifying underlying computational methods & theoretical foundations. Its effectiveness & usefulness are rigorously assessed.

DOI: 10.4018/979-8-3693-9934-7.ch004

1. INTRODUCTION

Computer programming is one of the very important aspects of our education, as it constitutes one of the fundamental concepts for interacting with computers. Given that computers are integrated in various sectors such as healthcare, transportation, defence, and our daily lives, understanding computer interactions will continue to proliferate in our daily routine, comprehending their functioning becomes crucial for tasks ranging from efficiency to repairs (Page, R. L., 2003). Therefore, computer programming holds paramount importance in our education. Moreover, even more critical than computer programming itself is the understanding of discrete mathematics. This foundation discipline serves as the bedrock for every computer program, software, or programming languages, as they all operate based on the fundamental principles of discrete mathematics (Henderson, 1990). Discrete mathematics, though a cornerstone of modern computing and mathematics, emerged gradually over centuries as a response to practical problems and theoretical inquiries. Its roots can be traced back to ancient civilizations, where early mathematicians began exploring concepts that would later form the basis of discrete mathematics. One of the earliest recorded instances of discrete mathematics can be found in the works of ancient Indian mathematicians, particularly in the field of combinatorics. Around the 6th century BCE, scholars like Pingala introduced the concept of binary numbers in the form of a poetic verse known as “Chandahśāstra”, which laid the foundation of modern binary arithmetic and combination analysis (Graham et al., 1995).

In ancient Greece, mathematicians such as Euclid and Pythagoras contributed to the development of discrete geometry and number theory. Euclid’s work on prime numbers and geometric proofs, as outlined in his famous work “Elements”, laid the groundwork for discrete mathematics concepts like number theory and graphs theory (O’Regan et al., 2016). This Islamic Golden Age (8th to 14th centuries CE) saw significant advancements on algebra and combinatorics by scholars like AI-Khawarizmi, AL-Kindi, and Omar Khayyam. Their contributions to algebraic equations, polynomial equations, and algebraic methods of solving combinatorial problems further enriched the field of discrete mathematics (Lovasz, 1993). During the Renaissance period, mathematicians such as Leonardo Fibonacci and Pierre de Fermat made notable contributions to number theory, particularly in areas like Fibonacci sequences, Diophantine equations, and modular arithmetic. These developments laid the groundwork for discrete mathematics as a distinct branch of mathematics (Alon & Spencer, 2000).

The 19th and early 20th centuries witnessed the formalization and expansion of discrete mathematics as a discipline in its own right. Mathematicians like Georg Cantor, Augustus De Morgan, and Richard Dedekind made significant contributions to set theory. Logic, and combinatorics, establishing the theoretical framework for modern discrete mathematics. In the mid-20th century, with the advent of digital computers and the need to solve discrete problems in computer science and cryptography, discrete mathematics gained newfound relevance and prominence. The development of algorithms, graph theory, combinatorial optimization, and discrete probability further solidified the importance of mathematics in various fields (Lovasz, 1993). Today, discrete mathematics plays a pivotal role in diverse areas such as computer science, cryptography, operations research, telecommunications, and data science.

Discrete Mathematics underpins many essential concepts and techniques in modern computer programming, serving as a foundational framework for solving a wide range of computational problems.

1.1. Algorithm Design and Analysis:

Discrete mathematics provides the theoretical foundation for designing and analysing algorithms. Concepts such as graph theory, combinatorics, and recurrence relations are instrumental in creating efficient algorithms for tasks such as sorting, searching, and optimization. For example, graph algorithms like Dijkstra's algorithm for shortest paths or dynamic programming techniques for optimization problems heavily rely on discrete mathematics principles (Rosen, 2019).

1.2. Data Structure:

Data structures are fundamental components of computer programming, enabling efficient storage and retrieval of data. Discrete mathematics plays a crucial role in the design and analysis of data structures such as arrays, linked lists, trees, heaps, and hash tables. Graph theory concepts are particularly relevant in representing and manipulating hierarchical or interconnected data structures (Rosen, 2019).

1.3. Cryptography:

Discrete mathematics forms the backbone of cryptographic algorithms used to secure data and communications in modern computing. Number theory, particularly modular arithmetic and prime factorization, is essential for developing encryption and decryption algorithms like RSA and Diffie-Hellman key exchange. Discrete logarithm problems and elliptic curve cryptography also rely heavily on discrete mathematical concepts (Rosen, 2019).

1.4. Combinatorial Optimization:

Many real-world problems involve optimizing a discrete set of choices subject to constraints. Discrete mathematics provides tools and techniques for solving combinatorial optimization problems efficiently. Applications include resource allocation, scheduling, network routing, and logistics optimization. Algorithms such as greedy algorithms, dynamic programming, and branch-and-bound methods are commonly employed in these domains (Rosen, 2019).

1.5. Graph Theory and Network Analysis:

Graph theory is a branch of discrete mathematics that deals with the study of graphs, which model pairwise relationships between objects. In computer programming, graph theory finds applications in network analysis, social network analysis, transportation networks, and computational biology. Graph algorithms like breadth-first search, depth-first search, and graph traversal techniques are used extensively in these domains (Aigner, 2007).

1.6. Automata Theory and Formal Languages:

Discrete mathematics concepts like finite automata, regular expressions, context-free grammar, and formal languages are foundational in theoretical computer science and compiler design. These concepts are crucial for understanding and implementing lexical analysis, parsing, and code generation in programming languages and compilers (Aigner, 2007).

1.7. Probability and Randomized Algorithms:

Discrete probability theory plays a vital role in the analysis of randomized algorithms and probabilistic data structures. Applications include randomized algorithms for sorting, searching, and approximation algorithms for NP-hard problems. Probabilistic data structures like Bloom filters and hash-based algorithms leverage discrete probability theory for efficient storage and retrieval of data in large-scale systems (Aigner, 2007).

2. CODE STRUCTURE:

The used Python program aims to handle set operations. Let's dive deep into its components:

2.1. Imports:

The program starts by importing modules like '**customtkinter**','**os**','**ctype**','**PIL**','**pywinstyles**' shown in Figure 1.

Figure 1. Imported modules

```
import customtkinter as ctk
import ctypes
import os
from PIL import Image
import pywinstyles
```

- '**Customtkinter**' is a front-end library used to develop modern GUI in '**tkinter**' style and syntax,
- '**pywinstyle**' use a python library for styling windows or GUI like opacity etc.
- '**Image in PIL**' Image library which exist inside of PIL library is used to import images in the script.
- '**ctypes**' ctypes access and provide information related to user's system, for example size of the screen.
- '**os**' os library is used in many ways, for example chaing the directory in the file or checking for specific file etc.

2.2. Working Directory:

The '**os.chdir()**' function is used to set the working directory to a specific path where the current working file are located, thip step is optional as some editor or even windows version starts the file in their respective directory shown in Figure.2.

Figure 2 Directory change to current working file directory

```
os.chdir(os.path.dirname(os.path.abspath(__file__)))
```

2.3. USER32 Screen Size:

User32 represent the current user and its system conFigureuration, as shown in Figure 3. We created a variable with name ‘user32’ which contain a function of ‘ctypes’, as previously mentioned ‘ctypes’ access system based information which can be access by the application, that’s what we have created here ‘ctypes.win11. user32’ access the system based information related the current user of the application, then ‘**user32.GetSysteMetrics(0)/(1)**’ retrieved the screen size of the user’s system as width and height respectively. Which is later used to create the fullscreen GUI, explained below.

Figure 3. USER32 for Screen Size

```
user32 = ctypes.windll.user32
screensize = user32.GetSystemMetrics(0), user32.GetSystemMetrics(1)
```

2.4. GUI ConFigureuration:

We have declared the library of ‘**customtkinter**’ as **‘ctk’**, so for our convenience we will be using ‘ctk’ in place of customtkinter, ‘ctk.set_appearance_mode.’ Set the appearance mode of the GUI/Application, it can be based on system’s preference or custom like light or dark mode. Then we set ‘ctk.set_default_color_theme(“dark-blue”)’ result is shown in Figure 4.

Figure 4. Title Theme of Application

A theme can be set as ‘dark-blue (current)’, ‘blue’, ’green’.

2.5. Class and Self:

We declared a class name APP with argument of ‘ctk.CTk’ this argument import the structure of customtkinter under with we will be creating and placing our GUI widgets shown in Figure 5.

Figure 5. Class declaration

```
class App(ctk.CTk):
    def __init__(self):
        super().__init__()
```

'__init__(self) function with super().__init__() auto initialize the function with its content while the word 'self' is argument pass with root name of the customtkinter application later. Basically this allows us to write functions in any order we desire, giving us an organised code shown in Figure 6.

Figure 6. Function Declaration Order

```
        submit_button=ctk.CTkButton(master=self, text="Submit",bg_color='#000001',corner_radius=10,command=self.func)
        submit_button.pack(pady=(50,0))
        pywinstyles.set_opacity(submit_button, color="#000001")

        result_label=ctk.CTkLabel(master=self, text='')

    def func(self):
        setA=set()
        setB=set()
```

Function Called Here

Function is Declared Here

2.6. Main GUI Page:

Here, we have written the whole Main page with its GUI, 'self.title("Equation Calculator")' function display the words 'Equation Calculator' in the title bar of the application, 'self.geometry(f"….")', takes the width and height previously stored in 'screensize' list and creates the application with those parameters.

Figure 7. Main page code

```
global set_A_entry,set_B_entry,result_label

self.title("Equation Calculator")
self.geometry(f"{screensize[0]}x{screensize[1]}+0+0")
background_image=ctk.CTkImage(light_image=Image.open("background_image.png"),
                              dark_image=Image.open("background_image.png"),
                              size=(screensize[0],screensize[1]))

background_image_label=ctk.CTkLabel(master=self, image=background_image, text="", width=screensize[0],height=screensize[1])
background_image_label.place(x=0,y=0)

set_A_entry=ctk.CTkEntry(master=self, placeholder_text="Enter set A", width=(screensize[0]/3),height=30,corner_radius=10,bg_color='#000001')
set_A_entry.pack(pady=(300,0))
pywinstyles.set_opacity(set_A_entry, color="#000001")

set_B_entry=ctk.CTkEntry(master=self, placeholder_text="Enter set B", width=(screensize[0]/3),height=30,corner_radius=10,bg_color='#000001')
set_B_entry.pack(pady=(50,0))
pywinstyles.set_opacity(set_B_entry, color="#000001")
```

‘CTkImage’ Customtkinter already have an image function which is used for storing and displaying images in the application but the normal image format like png and jpg is not supported by the function itself, that the reason why we have imported PIL/Image as shown in Figure 8. ‘Image.open’ function open a png formatted image and convert its format to image format that customtkinter supports. We have also declared the size of image as the screen of the application.

Figure 8. CTkImage

```
background_image=ctk.CTkImage(light_image=Image.open("background_image.png"),
                              dark_image=Image.open("background_image.png"),
                              size=(screensize[0],screensize[1]))
```

‘CTkImage’ cannot display the image as its whole, for this we have to use a widget as a base for the image, it can frame, button, labels etc. In our case we have used a ‘CTkLabel’, it’s an empty label with no text just an image as our image will used a background image we place the label at 0,0 coordinated (Note: coordinates are in pixels) which is the left topmost corner in your application shown in Figure 9.

Figure 9. Using labels as a base for images

```
background_image_label=ctk.CTkLabel(master=self, image=background_image, text="", width=screensize[0],height=screensize[1])
background_image_label.place(x=0,y=0)
```

Next we have set two entry boxes, as the name suggest it’s an entry box for taking user input, note what ‘pywinstyles’ is used to make the color ‘#000001’ invisible or set its opacity to 0, while the entry box’s background color is also set to ‘#000001’, the reason behind this is when we create a rounded corner in any widget, the first layer also known as fg follows the rules but the back layer known as bg does not, shown in Figure 10.

Figure 10. Differences in using of pywinstyle

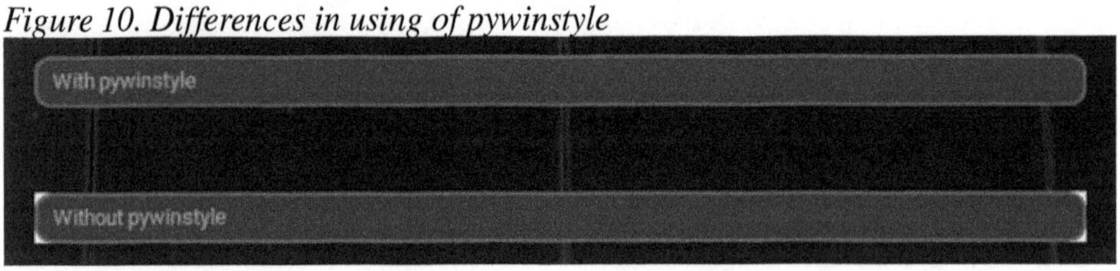

The combo function acts as a callback for the combo box (operation_combo_box). Whenever a user selects an operation (Union, Intersection, Difference, or Symmetric Difference) from the dropdown, this function is triggered. It takes the selected operation as an argument (choice) and assigns it to a global variable called operation. This ensures that the selected operation is globally accessible and can be used later when performing set operations in the func method shown in Figure 11.

Figure 11. Combo function

```
def combo(choice):
    global operation
    operation=choice
```

The ‘operation_combo_box’ is a dropdown menu created using the ‘CTkComboBox’ widget from the customtkinter library shown in Figure 12. It allows the user to select one of four operations—Union, Intersection, Difference, or Symmetric Difference. This combo box is conFigureured with several parameters. The values parameter provides the list of operations displayed in the dropdown. The command parameter links the combo box to the combo function, ensuring that the selected value is updated globally. Additionally, this widget also uses the ‘pywinstyles’ as shown above.

Figure 12. Operation combo box

```
operation_combo_box=ctk.CTkComboBox(master=self,values=["Union","Intersection","Difference","Symmetric Difference"],command=combo,bg_color='#000001')
operation_combo_box.pack(pady=(170,0))
operation_combo_box.set("Select")
pywinstyles.set_opacity(operation_combo_box, color="#000001")
```

The 'submit_button' is a button created using the CTkButton widget. This button is responsible for triggering the 'func' method when clicked. It is styled with a dark background color (#000001) for pywinstyle, rounded corners (corner_radius=10), and a text label (“Submit”). The command parameter is set to 'self.func', ensuring that the method responsible for performing the set operation is executed upon clicking the button. The button is placed below the combo box using the pack method with 50 pixels of vertical padding for spacing. The button serves as a crucial element in the workflow, allowing the user to process their input after selecting an operation shown in Figure 13.

Figure 13. Submit button

```
submit_button=ctk.CTkButton(master=self, text="Submit",bg_color='#000001',corner_radius=10,command=self.func)
submit_button.pack(pady=(50,0))
pywinstyles.set_opacity(submit_button, color="#000001")
```

The 'func' method is the core logic of the application, responsible for performing the selected set operation based on the user input. It begins by initializing two empty sets, setA and setB, and then clears any previously displayed result by hiding the result_label using the pack_forget method. Next, the method retrieves input from two entry fields (set_A_entry.get() and set_B_entry.get()) that the user is expected to fill. Each character from the input is added to the respective set, skipping commas. The method then checks the value of the global operation variable to determine which operation to perform. Using Python's built-in set methods, it calculates the result of Union, Intersection, Difference, or Symmetric Difference as per the selected operation. The result is then displayed by updating the text of result_label and making it visible on the window shown in Figure 14.

Figure 14. Func function

```
def func(self):
    setA=set()
    setB=set()

    result_label.pack_forget()

    for i in set_A_entry.get():
        if i == ',':
            continue
        else:
            setA.add(i)
    for i in set_B_entry.get():
        if i == ',':
            continue
        else:
            setB.add(i)
    if operation == 'Union':
        result_label.configure(text=setA.union(setB))
        result_label.pack(pady=(50,0))
    elif operation == 'Intersection':
        result_label.configure(text=setA.intersection(setB))
        result_label.pack(pady=(50,0))
    elif operation == "Difference":
        result_label.configure(text=setA.difference(setB))
        result_label.pack(pady=(50,0))
    elif operation == 'Symmetric Difference':
        result_label.configure(text=setA.symmetric_difference(setB))
        result_label.pack(pady=(50,0))
```

The main section of the script instantiates the App class, which represents the GUI application, and starts the main event loop using app.mainloop(). This ensures that the GUI remains responsive and interactive, allowing users to input data, select an operation, and view results dynamically shown in Figure 15.

Figure 15. Mainloop

```
if __name__=="__main__":
    app=App()
    app.mainloop()
```

2.7. GUI Result:

We have successfully completed the whole backhand and front-hand coding of the application; the final result is shown in Figure 16. Here we have two entry boxes in which a set of number separated by commas, then a combo box has operation choosing and finally a submit button, if pressed it creates a label below itself displaying the result of the operation performed on the two sets provided.

Figure 16. Result

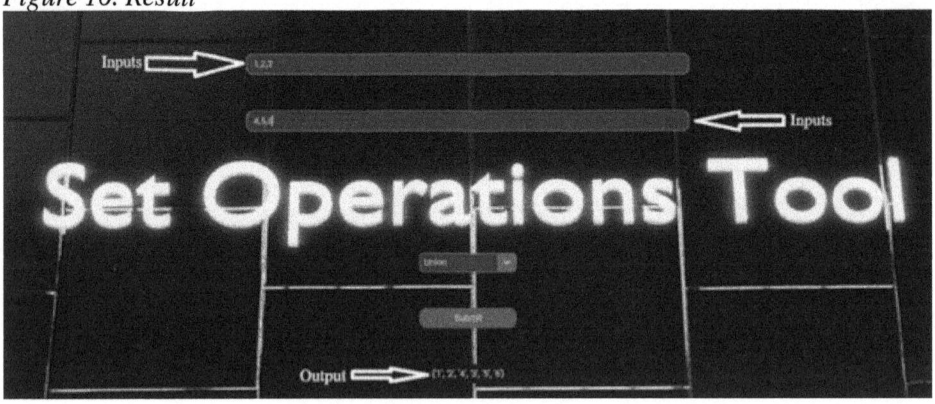

2.8. Set of Examples:

We have provided some inputs to our application to test it. Here are the examples.

Figure 17. First input

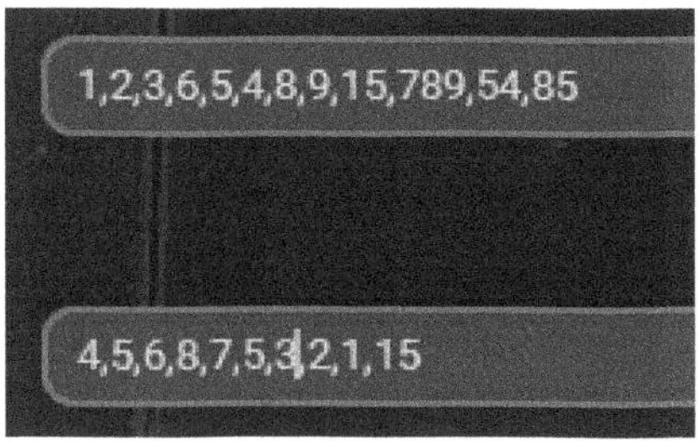

Figure 18. First output

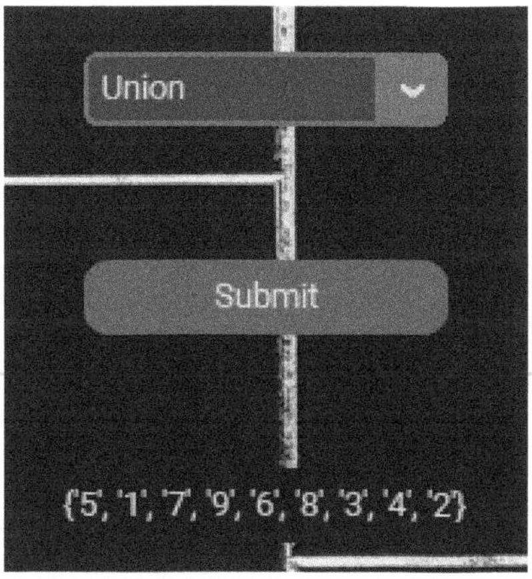

Figure 19. Second input

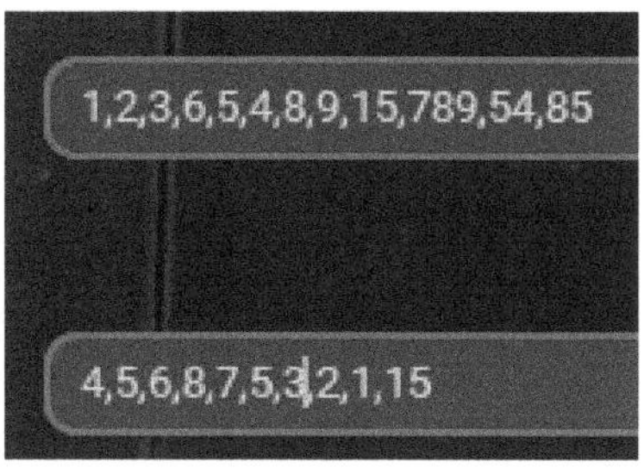

Figure 20. Second output

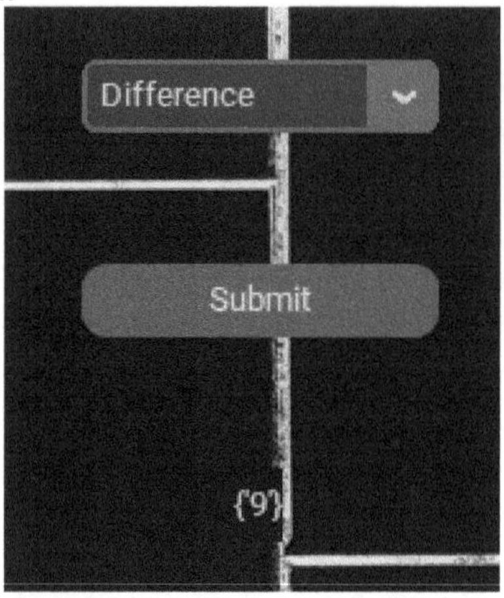

Figure 21. Third input

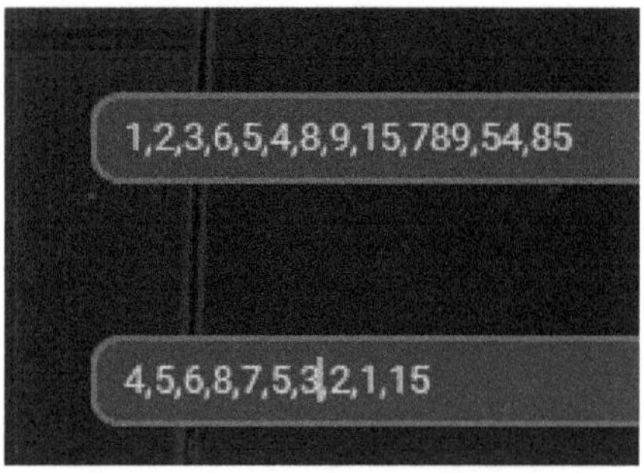

Figure 22. Third output

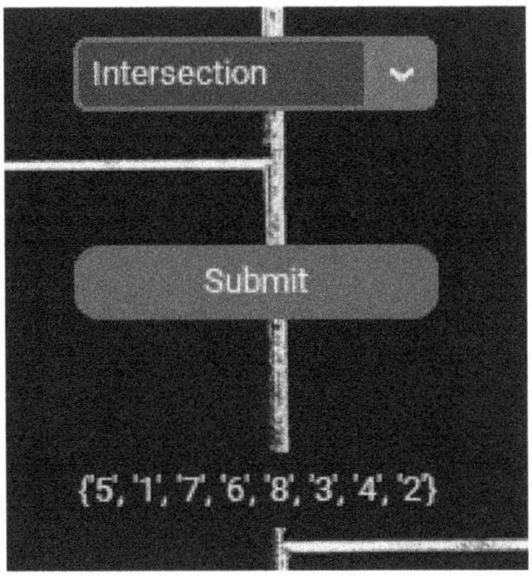

3. CONCLUSION

This report has successfully demonstrated the development and practical implementation of a comprehensive web-based software tool engineered using HTML, CSS, JavaScript, and Python to facilitate a broad spectrum of set operations. The use of Python, particularly in handling data storage and manipulation through CSV files, has significantly bolstered the tool's functionality, allowing for efficient data processing and user interaction. By leveraging the strengths of both client-side scripting (JavaScript for interactive web elements) and server-side scripting (Python for backend operations), the software provides a robust and user-friendly platform for educational purposes and practical applications in discrete mathematics.

The integration of Python has notably enhanced the versatility of the tool, enabling it to perform complex set operations such as union, intersection, difference, symmetric difference, power set generation, and Cartesian product computations with precision and ease. This not only demonstrates the power of web technologies in making mathematical tools more accessible but also highlights the role of Python as a pivotal technology in the development of educational software. The tool's effectiveness was rigorously evaluated through user studies, practical tests, and comparative analysis with existing tools, establishing its value in both educational settings and real-world problem-solving scenarios.

Future work will focus on expanding the capabilities of the software, possibly integrating more advanced mathematical operations and exploring the potential for mobile platform compatibility to enhance accessibility further. The continuous refinement of the software will aim to provide an even more intuitive user experience and deeper insight into the practical applications of discrete mathematics, driven by the seamless integration of Python and web technologies.

REFERENCE

Aigner, M. (2007). *Discrete Mathematics.* American Mathematical Society.

Alon, N., & Spencer, J. H. (2000). *The Probabilistic Method* (2nd ed.). Wiley. DOI: 10.1002/0471722154

Graham, R. L., Grotschel, M., & Lovasz, L. (1995). *Handbook Of Combinatorics.* North-Holland.

Henderson, P. B. (1990). *Discrete mathematics as a precursor to programming.* ACM SIGCSE Bulletin., DOI: 10.1145/323410.319070

Lovasz, L. (1993). *Combinatorial Problems and Exercises* (2nd ed.). North-Holland.

O'Regan, G., Gries, D., & Schneider, F. B. (2016). *Guide to Discrete Mathematics.* Springer. DOI: 10.1007/978-3-319-44561-8

Page, R. L. (2003). *Software is discrete mathematics.* ACM SIGPLAN International. DOI: 10.1145/944705.944713

Rosen, K. H. (2019). *Discrete Mathematics and Its Applications* (8th ed.). McGraw-Hill.

Chapter 5
Representations in Developing Statistical Thinking in Secondary Education

P. Selvakumar
https://orcid.org/0000-0002-3650-4548
Department of Science and Humanities, Nehru Institute of Technology, Coimbatore, India

Ashwarya Sharma
https://orcid.org/0009-0005-8510-3862
Chandigarh School of Business, Chandigarh Group of Colleges, Jhanjeri, India

Sourabh Singh Roy
Gopal Narayan Singh University, India

Mohit Sharma
https://orcid.org/0009-0007-2280-8077
Maharshi Dayanand University, Rohtak, India

T. C. Manjunath
https://orcid.org/0000-0003-2545-9160
Rajarajeswari College of Engineering, India

ABSTRACT

The chapter highlights the critical role that representations play in facilitating a deeper grasp of statistical thinking, helping students not only perform statistical calculations but also interpret and apply statistical findings in real-world contexts.

DOI: 10.4018/979-8-3693-9934-7.ch005

At its core, the chapter emphasizes the idea that statistical reasoning is not just about learning formulas and procedures but involves the ability to represent data and concepts in various ways that enhance understanding and improve problem-solving. Statistical education is inherently about managing uncertainty, recognizing patterns, and empiricism. However, students often find it difficult to grasp abstract statistical concepts, as they can seem disconnected from real-world experience. By focusing on the use of representations, the chapter suggests providing more tangible ways to engage with complex topics. One of the primary goals of this chapter is to demonstrate how representations act as cognitive tools that aid in transforming abstract ideas into more accessible, understandable forms.

INTRODUCTION TO REPRESENTATIONS IN STATISTICAL EDUCATION: OVERVIEW OF THE CHAPTER AND ITS SIGNIFICANCE

The chapter highlights the critical role that representations play in facilitating a deeper grasp of statistical thinking, helping students not only perform statistical calculations but also interpret and apply statistical findings in real-world contexts. At its core, the chapter emphasizes the idea that statistical reasoning is not just about learning formulas and procedures but involves the ability to represent data and concepts in various ways that enhance understanding and improve problem-solving. Statistical education is inherently about managing uncertainty, recognizing patterns, and empirical. However, students often find it difficult to grasp abstract statistical concepts, as they can seem disconnected from real-world experience. By focusing on the use of representations, the chapter suggests providing more tangible ways to engage with complex topics. One of the primary goals of this chapter is to demonstrate how representations act as cognitive tools that aid in transforming abstract ideas into more accessible, understandable forms. For example, statistical graphs, such as histograms, allow students to distribute relationships between variables. In addition, symbolic representations, such as mathematical equations and formulas, offer a precise way to express statistical concepts, while verbal representations facilitate clear communication of statistical ideas. Interactive tools, such as simulation software and data visualization platforms, provide students with opportunities to explore data dynamically, reinforcing their understanding through hands-on experience.

Furthermore, the chapter acknowledges the challenges faced by both students and educators in statistical education. One of the central challenges is the abstract nature of many statistical concepts. Students often struggle to connect theoretical concepts with real-world data, and educators must find ways to bridge this gap. The chapter highlights how effective representations can serve as cognitive scaffolding, supporting students as they navigate complex statistical problems. By providing students with opportunities data, educators help them develop the skills needed to approach statistical challenges with confidence. The chapter also addresses the importance of technology in modern statistical education. In the digital age, students have access to tools that can represent data. From statistical software packages to online platforms for interactive learning, technology offers new opportunities for enhancing statistical education. The chapter explores how these tools can be integrated into the classroom to provide students with hands-on experiences and immediate feedback. For example, online simulations and data visualization platforms allow students to explore statistical concepts in a dynamic, interactive environment, helping them develop a more robust understanding of statistical principles. Moreover, the chapter explores the significance of developing statistical literacy—an essential skill for the 21st century. Statistical literacy goes beyond the ability to perform basic calculations; it encompasses the ability to critically evaluate data, interpret statistical results, and make informed decisions based on evidence. Representations developing statistical literacy students engage with data in meaningful ways. By using visual representations to demonstrate the effects of different statistical methods, educators can help students build a stronger foundation in statistical reasoning. Finally, the chapter recognizes that effective statistical education requires more than just the use of representations; it also requires an understanding of how to integrate these representations into a coherent teaching strategy. The chapter provides practical advice for educators on how to use representations in the classroom, offering strategies for incorporating them into lectures, assignments, and assessments. It also highlights the importance of active learning, encouraging educators to create opportunities for statistical concepts in an interactive manner.

THE IMPORTANCE OF STATISTICAL LITERACY IN SECONDARY EDUCATION: DISCUSSION OF THE ROLE OF STATISTICAL LITERACY IN SECONDARY EDUCATION

Statistical literacy has become an essential component of secondary education as the world becomes increasingly data-driven. In an era where decisions, both personal and societal, are often influenced by data and statistical findings, the ability to critically interpret, understand, and apply statistics is more important than ever. Statistical literacy refers to the knowledge and skills required to interpret and analyze data, understand probability, and make informed decisions based on statistical reasoning. It empowers students to engage with the world around them, enabling them to critically evaluate data in various contexts, from media reports to scientific studies. At the heart of statistical literacy is the understanding that data is not just numbers, but a reflection of real-world phenomena. In secondary education, students are introduced to various statistical concepts of probability. Concepts serve as the building blocks for more advanced study of science, psychology, and studies (Laupichler M C., et al, 2022). In addition, practical application of these concepts, students learn the importance of ethical data handling and the potential for misuse or manipulation of statistical information. The ability to critically assess sources of data, the methods used to collect it, and the validity of conclusions drawn from it is a vital skill in a world where misinformation is rampant. Statistical literacy also encourages develop a mindset for scientific inquiry. By engaging in real-world problems, students learn how to collect data. For example, students might work on projects where they analyze survey data or conduct experiments, using statistical tools to summarize findings, identify patterns, and test hypotheses. These hands-on activities not only reinforce the theoretical concepts learned in the classroom but also demonstrate the power of statistics in uncovering insights about the world. Furthermore, statistical literacy fosters critical thinking. In an age where large volumes of data are accessible through digital platforms, students must develop the ability to distinguish between reliable and unreliable data sources. They must also understand the limitations of statistical analysis, such as the potential for bias in sampling or the misinterpretation of statistical measures (Nikou S., et al, 2021). By developing these skills, students are better equipped to make decisions based on evidence, whether they are choosing the best course of action in a personal situation or evaluating the effectiveness of public policies. The role of statistical literacy in helping students navigate the complexities of the information age cannot be overstated.

Moreover, technology plays an increasingly significant role in the availability of software and online tools that have made us engage with and perform statistical analysis. Programs like Excel, Google Sheets, and specialized statistical software enable students to manipulate data sets, visualize trends, and apply statistical methods to conclude. The integration of technology into the curriculum and understanding of statistics prepares them for the tools they will encounter in higher education and the workforce (Sitopu J W., et al, 2024). Despite its importance, there are several challenges to promoting statistical literacy in secondary education. One of the primary challenges is the lack of adequate training for teachers. To address this issue, educators must adopt differentiated instruction strategies, such as scaffolding complex ideas and providing additional support to struggling learners. By fostering an inclusive learning environment, schools can help all students develop the statistical literacy they need to succeed in an increasingly data-driven world (Susilawati E., et al, 2021).

CHALLENGES IN TEACHING AND LEARNING STATISTICS IN SECONDARY EDUCATION: ANALYSIS OF THE CHALLENGES FACED BY STUDENTS AND INSTRUCTORS IN STATISTICS EDUCATION

In addressing these challenges, it is important to consider both the difficulties faced by students in comprehending statistical concepts and the obstacles instructors encounter when attempting to teach the subject effectively. One of the primary challenges in teaching statistics is the inherent difficulty of the subject itself. Unlike many other areas of mathematics, statistics involves not only numerical computation but also the interpretation of data, understanding variability, and drawing conclusions based on probability. For many secondary school students, this shift from purely mathematical processes to the application of mathematical concepts to real-world data can be a significant hurdle (Kamal T., et al, 2021). The abstract nature of probability, the complexities of distributions, and the nuances of sampling can confuse students who may struggle to see the real-world applications of these ideas. This often results in a sense of alienation from the subject, with students perceiving statistics as an irrelevant or overly technical field. In addition to the conceptual challenges, students often struggle with the mathematical foundations required for a deep understanding of statistics. Many students entering secondary education may not have mastered fundamental concepts such as algebra or basic probability theory, which are crucial to understanding more advanced statistical ideas. Without a solid grasp of these foundational concepts, students may find it difficult to engage with more complex statistical topics, such as hypothesis testing, regression analysis, or analysis of variance. This knowledge gap can hinder the learning process, as students may struggle to connect new statistical concepts to their existing mathematical understanding. Another challenge is the diversity of student backgrounds and abilities. Secondary school classrooms often contain students with varying levels of proficiency in mathematics and statistics (Mutohhari F., et al, 2021).

Many educators, especially those in secondary schools, may not have specialized training in statistics. As a result, they may lack the depth of knowledge required to effectively teach the subject, particularly in areas such as statistical inference, probability theory, and experimental design (Zou C., et al, 2021). Even teachers with a strong background in mathematics may find it difficult to translate their knowledge into effective teaching strategies for a subject that requires both conceptual understanding and practical application. Furthermore, statistics is often treated as a subdiscipline within broader mathematics curricula, rather than as a standalone subject, which can lead to a lack of emphasis on the importance of statistical literacy. Instructors face the challenge of designing assessments that accurately measure students' statistical literacy and statistical. Beyond the classroom, external factors can also impact the teaching and learning of statistics. Standardized testing, for example, often emphasizes content areas such as algebra and calculus, with little regard for statistical reasoning or data analysis (Basar Z M., et al, 2021). As a result, teachers may feel pressured to prioritize content that aligns with standardized test expectations, neglecting the importance of statistical literacy. May inadvertently contribute to the marginalization of statistics, as it is often viewed as a subset of mathematics rather than a critical discipline in its own right. Overcoming these external pressures requires advocacy for the inclusion of statistics as a core subject within educational frameworks and a recognition of its importance in developing well-rounded, data-literate students.

VISUAL REPRESENTATIONS OF STATISTICAL CONCEPTS: GRAPHS, CHARTS, AND PLOTS: EXPLORATION OF VISUAL REPRESENTATIONS OF STATISTICAL CONCEPTS

Visual representations of statistical concepts, such as graphs, charts, and plots, are essential tools in helping students, researchers, and professionals understand and interpret data. These visual tools provide a clear and intuitive way to present complex statistical information, enabling individuals to discern patterns, relationships, and trends that might be difficult to identify in raw numerical data alone. The importance of effective data visualization extends beyond the classroom; in a world where vast amounts of data are generated every day, the ability to communicate data insights visually has become a vital skill in numerous fields, including economics, medicine, business, and social sciences. By examining the various types of graphs, charts, and plots used to represent statistical concepts, we can better understand how they facilitate the communication of data, enhance statistical literacy, and promote critical thinking (Yau N., 2024). One of the most fundamental types of visual representation is the bar chart, which is used to display categorical data. Bar charts allow for a quick comparison of the frequency or proportion of different values. Bar charts are particularly useful when comparing discrete categories, such as the performance of different groups in an experiment. A key advantage of bar charts is their simplicity and ease of interpretation, making them a staple in introductory statistics education. However, users need to understand the limitations of bar charts, such as the potential for distortion when the axis is not scaled properly or when bars are arranged in a misleading order. Another common type of chart is the pie chart, which is often proportional categorical pie chart divides a circle into segments, with each segment representing a category and its size proportional to the value it represents. While pie charts can be visually appealing and easy to understand, they are not always the most effective way to compare categories, particularly when there are many small differences between the segments. Additionally, pie charts can become cluttered and difficult to interpret if there are too many categories, which can obscure the data's true patterns (Franconeri S L., et al, 2021). Despite these limitations, pie charts are widely used in various fields, from marketing and business to education, to provide a snapshot of data distribution in an easily digestible format.

A positive or negative slope of the line indicates a positive or negative correlation, respectively. However, it is important to exercise caution when interpreting scatter plots, as correlations do not imply causation and other underlying factors may influence the relationship between the variables. Line graphs are another commonly used over time. Line graphs connect These graphs, such as changes, or Line graphs provide a clear representation of how a variable changes over time and allow for easy comparisons between multiple series (Farahani A M., et al, 2023). However, like scatter plots, line graphs can be misleading if the scale is manipulated or if too many variables are plotted on the same graph, making it difficult to discern individual trends. Line graphs are most effective when the goal is to illustrate temporal patterns or compare multiple trends over the same period. While these visual tools are essential in the representation of statistical concepts, it is important to recognize that no single visualization is universally best for all types of data. The choice of graph, chart, or plot being presented, the audience's familiarity with statistical concepts, and the specific goals of the analysis. Overcomplicating a visualization with excessive decoration or unnecessary elements can detract from its clarity, making it harder for the audience to extract meaningful insights from the data. Furthermore, the design of a graph or chart should be guided by the audience's level of statistical literacy, with careful consideration given to how much background knowledge the audience is likely to have (Xu H., et al, 2022). For example, a line graph for a general audience should avoid overly technical terms and focus on communicating the main trends, while a more technical audience may appreciate a more detailed breakdown of the data.

SYMBOLIC REPRESENTATIONS OF STATISTICAL CONCEPTS: FORMULAS AND EQUATIONS: DISCUSSION OF SYMBOLIC REPRESENTATIONS OF STATISTICAL CONCEPTS

Symbolic representations, such as formulas and equations, and applications serve as the language through which complex statistical concepts are formalized and communicated. These symbolic forms allow for a precise and compact expression of relationships between variables, the calculation of important statistics, and the derivation of conclusions from data. Whether in the context of descriptive statistics, inferential statistics, or advanced statistical models, formulas and equations provide a systematic way to describe patterns, quantify uncertainty, and make predictions (Beranger B., et al, 2023). Where, is size, and represents formula allows for formal expression providing critical insight into the variability of the data. In addition to central tendency and variability, symbolic representations also extend to measures of the distribution's shape (Kronberger G., et al, 2024). Sample size The SEM is an essential concept in hypothesis testing and the construction of confidence intervals, as it provides accurate an estimate. By incorporating the symbolic representation of the SEM, statisticians can derive critical values for various statistical tests, such as the t-test, which are used to make inferences about population parameters. While symbolic representations are indispensable for formalizing statistical concepts and conducting analyses, they can also present challenges, particularly for students who are new to statistics. The abstract nature of formulas and equations requires a solid understanding of mathematical principles, as well as the ability to interpret and manipulate symbols to solve problems. For students unfamiliar with statistical notation, formulas can initially seem daunting and unintuitive (Angelis D., et al, 2023). However, with careful instruction and practice, students can learn to navigate these symbolic representations, gaining a deeper understanding of statistical concepts and their applications.

VERBAL REPRESENTATIONS OF STATISTICAL CONCEPTS: LANGUAGE AND COMMUNICATION: ANALYSIS OF VERBAL REPRESENTATIONS OF STATISTICAL CONCEPTS

Verbal representations of statistical concepts are essential for conveying complex ideas to a wide range of audiences, from students and researchers to policymakers and the general public. Unlike symbolic representations, which use mathematical notation to express statistical relationships, verbal representations use language to describe, explain, and interpret statistical phenomena. Language centralizes both teaching and helps bridge the gap between technical statistical formulas and real-world contexts. Verbal representations of statistical concepts are particularly crucial in making statistical reasoning more intuitive, providing clarity, and enabling individuals to communicate statistical findings effectively. However, the challenge lies in ensuring that these verbal descriptions are accurate, clear, and appropriate for the target audience's level of understanding. One of the most fundamental verbal representations in statistics is the concept of descriptive statistics, which involves summarizing and organizing data to identify patterns and trends. Descriptive statistics are often communicated verbally through statements such as “the average score of the class was 75,” or “the distribution of incomes shows a significant skew to the right.” These verbal descriptions are crucial for translating raw data into comprehensible information that can be easily communicated to non-experts (Kiefer M., et al, 2022). On the other hand, terms like measure squared deviation from mean square root variance, respectively, are more abstract and require more precise verbal elaboration (Gal I., et al, 2022). In everyday language, one might say, “The data are spread out a lot,” but in statistical terms, the explanation would involve a more nuanced description of the distribution’s spread and the calculation behind the measure. Beyond descriptive statistics, verbal representations are also essential when communicating a In inferential statistics, verbal explanations often focus on the concepts of probability, For example, when describing the results of a hypothesis test, one might say, “There is strong evidence that the new drug is more effective than the old one,” or “The p-value of 0.03 indicates that there is a 3% chance of observing this result if the null hypothesis were true.” These verbal descriptions must be used carefully, as misinterpretations of terms like “p-value” or “confidence level” can lead to confusion. In particular, the p-value is often misunderstood as, when it is Similarly, confidence intervals are sometimes misinterpreted as in will always, when they represent in fall, given based on sample data.

In addition to describing specific statistical concepts, verbal representations are crucial in explaining the meaning of statistical results in the context of real-world applications. For instance, when interpreting the results of a regression analysis, one might say, “For every one-unit increase in the independent variable, the dependent variable increases by 2 units, on average.” Here, verbal communication helps to translate the mathematical output of the regression model into an understandable form, enabling the audience to grasp the practical implications of the analysis (Tong D H., et al, 2021). Communicating these assumptions is crucial for ensuring that the audience understands the conditions under which the statistical model is valid. Similarly, when discussing the limitations of a study, verbal representations help clarify how the sample size, study design, or potential biases may affect the reliability and generalizability of the results. The need for clarity and precision in verbal representations becomes even more pronounced when communicating statistical findings to non-experts or policymakers (Boyd R L., et al, 2021). In these cases, statistical terminology must be carefully tailored to the audience’s level of understanding, and the use of jargon or overly technical language should be minimized.

DEVELOPING STATISTICAL THINKING: FROM DESCRIPTIVE TO INFERENTIAL STATISTICS: DISCUSSION OF THE DEVELOPMENT OF STATISTICAL THINKING IN STUDENTS

Developing statistical thinking in students is a critical process that bridges the gap between understanding basic concepts and applying statistical methods to solve real-world problems. The evolution of statistical thinking involves a progression from a foundational understanding of descriptive statistics to the more advanced concepts of inferential statistics. This development equips students with the ability to not only describe and summarize data but also to make informed decisions, draw valid conclusions, and understand the underlying uncertainty associated with those conclusions. Statistical thinking goes beyond just learning formulas and procedures; it fosters a mindset where students approach problems with a critical understanding of data, probability, and variability, and where they can communicate statistical results in meaningful and practical ways. By emphasizing the development of statistical thinking from descriptive to inferential statistics, educators can help students build a solid foundation for understanding complex data analysis, which is essential in fields ranging from economics and psychology to healthcare and business (Estrella S., et al, 2022). The journey toward statistical thinking typically begins with an introduction to descriptive statistics, which focuses on summarizing and organizing data to make it more comprehensible. At this stage, students are introduced to basic. These concepts are crucial as they provide the first steps in recognizing patterns in data. Descriptive statistics form the foundation for statistical thinking by enabling students to summarize large datasets in a meaningful way. For example, tendency, while standard deviation quantifies the spread or variability in the data. Rather than simply summarizing the data, they now use sample estimate parameters. The idea of sampling distributions describes how they vary from one sample to another. This process involves statistics (Nusantara T., et al, 2021).

Similarly, ANOVA extends the concept of hypothesis testing by comparing means across multiple groups, allowing students to test and explore factors that might be driving those differences (Koga S., 2022). Developing statistical thinking is not only about acquiring technical knowledge; it also involves fostering a mindset of critical inquiry and skepticism. Students must be encouraged to think about the limitations of statistical methods, including the assumptions that underpin statistical models and the potential for bias or error in data collection. For example, when conducting a survey or experiment, students must consider the design of the study, including issues such as randomization, sample size, and potential sources of bias. Effective communication also involves the ability to visualize data in ways that highlight key patterns and trends and to make decisions based on statistical evidence. Developing these communication skills is crucial for students who will go on to use statistics in fields such as business, healthcare, education, and public policy (Van Dijke-Droogers M., et al, 2022). The development of statistical thinking practice, and continual engagement with statistical concepts. From the initial stages of learning descriptive statistics to the more advanced application of inferential methods, students must be provided with opportunities to engage with real-world data, apply statistical techniques, and critically assess the validity and relevance of their findings.

COGNITIVE BIASES AND HEURISTICS IN STATISTICAL REASONING: ANALYSIS OF COGNITIVE BIASES AND HEURISTICS THAT INFLUENCE STATISTICAL REASONING

Cognitive biases and heuristics play a significant role in shaping how individuals' reason about statistics, often leading to systematic deviations from rational statistical thinking. Shortcuts especially when time or information is limited, and misinterpretation of data. Are inherent in human cognition and influence how people interpret, process, and evaluate statistical information. Arise statistical reasoning, which, while efficient, can sometimes result in incorrect or irrational conclusions. Understanding the various cognitive biases and heuristics that affect statistical reasoning is essential for both students and practitioners of statistics, as awareness of these biases can help mitigate their impact and improve decision-making processes. Well-known statistical reasoning is the availability heuristic. Heuristic individuals are the most readily available information to make judgments, often giving undue weight to information that is recent, memorable, or emotionally charged. For example, when evaluating the likelihood of a plane crash, an individual may overestimate the risk based on media coverage of recent high-profile accidents, even though statistical data shows that flying is much safer than driving (Martín M., et al, 2023). This bias is particularly problematic in the interpretation of probability, people's anecdotal evidence or vivid examples rather than statistical broader availability heuristic led to overestimations of rare events and underestimations of more common occurrences, distorting statistical reasoning and decision-making. Another common bias in statistical reasoning is the anchoring effect, where “anchor” occurs even when the anchor is arbitrary or irrelevant. This bias can result in skewed interpretations of data, as individuals fail to consider the full range of evidence, which is essential for making objective and accurate statistical inferences (Friedman H H., 2023). Confirmation bias can also affect how people evaluate the significance of statistical results, leading them to misinterpret p-values or confidence intervals. Their representativeness heuristic is another cognitive shortcut that influences statistical reasoning, particularly when individuals make judgments about probabilities or likelihoods resembles This often leads individuals to make erroneous conclusions because they rely on superficial characteristics rather than statistical information.

In statistical reasoning, overconfidence can manifest in the belief that one has a better grasp of statistical concepts than is the case, leading to the misapplication of statistical techniques or an underestimation of the uncertainty associated with statistical results (Šrol J., 2022). For example, a person might believe that the outcome randomized experiment with a small sample fails to account for the variability inherent. Overconfidence in judgment, such as failing to adjust conclusions based on new or contradictory evidence or relying too heavily on point estimates without considering the variability captured by confidence intervals. Another important cognitive bias is the hindsight bias, often referred to as the "knew-it-all-along effect." This bias occurs when individuals or obvious, even though there was no way to predict it with certainty beforehand. In statistical reasoning, and hindsight people misinterpret the past and overestimate the accuracy of their predictions. Similarly, when interpreting statistical findings in organizational contexts, managers may attribute positive results to their leadership or decision-making while blaming failures on external factors or circumstances beyond their control (Kliegr T., et al, 2021). Attribution bias can skew interpretations of statistical evidence, leading to overgeneralizations or overly simplistic conclusions about causality. The influence of cognitive biases and heuristics on statistical reasoning highlights the importance of promoting awareness and developing strategies One approach is to teach students and practitioners to think critically about statistical evidence and to encourage skepticism when interpreting data.

TEACHING STRATEGIES FOR PROMOTING DEEP UNDERSTANDING OF STATISTICAL CONCEPTS: EXPLORATION OF TEACHING STRATEGIES FOR PROMOTING DEEP UNDERSTANDING OF STATISTICAL CONCEPTS

Teaching statistical concepts effectively requires more than just delivering content through lectures or textchapter exercises. It demands active engagement, the use of varied teaching strategies, and the creation of learning environments where students can develop deep statistical concepts that go beyond the ability to memorize formulas or solve problems using standard procedures. It involves thinking critically about data, inferences communicating statistical findings clearly and accurately. To promote deep understanding, educators must adopt teaching strategies that emphasize active learning, foster conceptual understanding, and encourage students to apply statistical principles to real-world contexts. One of the most effective strategies for promoting a deep understanding of statistical concepts is to require them to apply statistical concepts, collaborate with peers, and reflect on their learning. Examples of active learning in statistics include problem-based learning (PBL), group projects, data analysis exercises, and case studies. In problem-based learning, students are presented with complex, open-ended problems that require and interpret. These problems may be drawn from real-world scenarios, such as analyzing the impact of a public health intervention or evaluating the effectiveness of a marketing campaign (Harefa D H., et al, 2024). By working with data, students also encounter the challenges of real-world data, such as missing values, outliers, and measurement errors, which encourages them to think critically about data quality and the limitations of statistical methods. For example, when teaching regression analysis, instructors can use scatterplots to show and illustrate how Visual tools also distribute to highlight the concept of sampling variability in statistical inference (Ruiz-Rojas L I., et al, 2023). By incorporating visualization into the teaching of statistics, educators provide students with a means of seeing patterns and trends in data that might otherwise be obscured by numerical summaries alone.

To support conceptual understanding, it is essential to connect statistical concepts to real-world applications. When students understand the practical significance of statistical methods, they are more likely to see the relevance of what they are learning and be motivated to engage with the material. Real-world applications of statistics can be drawn from healthcare, social sciences, business, and environmental studies. When teaching hypothesis testing, instructors can present students with a case study in which they must determine whether a new medical treatment is effective based on clinical trial data. By connecting statistical concepts to tangible outcomes, students can see how statistical methods are used in the real world. Additionally, instructors can highlight the ethical considerations involved in the use of statistics, such as how data can be misinterpreted or manipulated to support biased conclusions. This approach helps students appreciate the importance of statistical literacy in a wide range of fields and encourages them to use statistical methods responsibly. Another important strategy for promoting a deep understanding of statistics is scaffolding instruction. Scaffolding involves providing students with structured support as they progress through increasingly complex concepts, gradually removing that support as they become more confident and capable. In statistics, scaffolding can take many forms, such as breaking down complex topics into smaller, more manageable parts, providing step-by-step guides for solving problems, and offering opportunities for guided practice (Alam A., 2023). For example, when introducing inferential statistics, an instructor might first focus on simple concepts like confidence intervals and hypothesis testing before moving on to more advanced topics like regression analysis or analysis of variance (ANOVA). During this process, instructors can offer feedback and clarifications to help students build their understanding incrementally. Scaffolding also encourages metacognition, as tracks their progress over time.

This approach helps with how statistical methods fit together and how they can be applied to a wide variety of problems. Formative assessment is another key strategy for promoting deep understanding in statistics education. Formative assessments provide instructors feedback be struggling. By working together, students can learn from each other's strengths, ask questions in a non-threatening environment, and reinforce their understanding of the material. Peer teaching is another powerful strategy, where students take on the role of the instructor and explain concepts to one another (Kumar D., et al, 2023).

USING STATISTICAL SOFTWARE TO REPRESENT STATISTICAL CONCEPTS: DISCUSSION OF THE USE OF STATISTICAL SOFTWARE TO REPRESENT STATISTICAL CONCEPTS

The ability to use statistical software effectively allows students to move beyond theoretical knowledge of statistical concepts and problems, making the study of statistics more engaging, accessible, and relevant. One of the most powerful features of statistical software is its ability to visualize data in various forms, such as histograms, boxplots, scatterplots, and bar charts. These visual representations are invaluable for helping students understand fundamental concepts of distribution shapes, in particular, visualizations can help students concretely grasp abstract ideas by allowing them to see the data in ways that are easier to interpret than raw numbers or tables (Mertler C A., et al, 2021). For example, when learning about the distribution of a dataset, students can use a histogram to observe the shape of the distribution and identify features such as skewness, modality, and outliers. In doing so, they can gain an intuitive understanding of key concepts like normality and the central limit theorem. By running simulations, students can also experiment with statistical techniques such as hypothesis testing and confidence intervals, deepening their understanding of how these methods work in practice (Hosseini M., et al, 2024). Furthermore, simulations can be used to teach statistical concepts that might be difficult or impractical to demonstrate with real-world data, such as rare events or extreme values. Another significant advantage of using statistical software is the ability to perform complex statistical analyses that would otherwise be cumbersome or infeasible by hand.

In addition to these practical advantages, statistical software provides an opportunity for teaching programming skills, particularly with tools like R, Python, and SAS, which require students to write code to execute analyses. Learning to program in a statistical software environment allows students to develop computational skills that are increasingly valuable in the modern data-driven world. Programming in R or Python, for example, teaches students how to manipulate data, automate repetitive tasks, and perform custom analyses, all of which are highly transferable skills in fields ranging from data science to economics and healthcare (Ringle C M., et al, 2023). By incorporating programming into statistical instruction, educators can prepare students for the demands of the job market, where proficiency in statistical software is often a prerequisite for positions in research, data analysis, and analytics. In this way, the use of statistical software not only enhances students' understanding of statistics but also equips them with essential technical skills. Statistics using software has the potential to focus too heavily on the mechanics of using the software, rather than on understanding the underlying statistical concepts. To address this issue, instructors must provide clear guidance on how to use statistical software as a tool to support, rather than replace, conceptual understanding. For instance, when teaching hypothesis testing, instructors can first explain the theoretical framework behind hypothesis testing and then demonstrate how to perform the test using statistical software. For example, when teaching regression analysis, students should be taught how to interpret the regression coefficients, and assess, and understand the significance of p-values. By focusing on interpretation skills, instructors can help students bridge the gap between the software output and the statistical concepts it represents (Memon M A., et al, 2021). Moreover, using statistical software to teach statistics provides the opportunity to engage students with real-world data, which is often messy, and complex, and requires students to make decisions about how to clean, organize, and analyze the data.

INTERACTIVE VISUALIZATIONS OF STATISTICAL CONCEPTS: DYNAMIC GRAPHS AND CHARTS: EXPLORATION OF INTERACTIVE VISUALIZATIONS OF STATISTICAL CONCEPTS

Similarly, in a regression analysis, students can adjust the values of predictor variables to see how the regression line changes, providing a more intuitive understanding of concepts like slope, and intercept, By interacting with these dynamic graphs, students can visualize how statistical models behave under different conditions, which helps them internalize the underlying concepts more effectively (Yau N., 2024). Interactive visualizations also facilitate the exploration of multivariate data in a way that is difficult to achieve with static visualizations. For example, when examining data with multiple variables, interactive tools such as 3D scatterplots, parallel coordinate plots, or interactive heatmaps allow students three simultaneously. By interacting with these visualizations, students can rotate the axes, zoom in on specific data points, and filter the data to examine particular subsets or trends. This level of interactivity helps students grasp the complexity of multivariate relationships and better understand concepts like multicollinearity, interactions, and higher-order relationships. For instance, in the context of a dataset examining factors influencing housing prices, students can use an interactive visualization to explore how a variable's location interacts to affect the price of a house (Steed C A., 2025). Interactive visualizations help bring these concepts to life by providing a visual representation of how data behaves in practice (Ferster B., 2023). For example, students can use an interactive probability distribution tool to explore normal, binomial, and Poisson and adjust the parameters (e.g., mean, variance, sample size) to see how the distribution changes. This allows students to visualize the differences between distributions and understand the properties of each one in a way that is difficult to convey through text or static diagrams alone. Additionally, interactive tools for hypothesis testing can help students visualize the p-value and understand its significance by simulating the process of collecting, calculating test statistics, and comparing it to a critical value. In addition, by discussing their analyses with others, students can clarify their understanding of statistical concepts and reinforce their learning through teaching. Instructors also benefit from using interactive visualizations in the classroom. These tools allow instructors to present complex concepts engagingly and interactively, making abstract ideas more tangible and accessible (Burch M., et al, 2021). For example, instructors can use interactive tools to demonstrate how changing the parameters of a statistical model affects the results, or they can guide a series of interactive visualization steps involved in hypothesis testing or regression analysis.

ONLINE RESOURCES AND TOOLS FOR TEACHING STATISTICAL CONCEPTS: ANALYSIS OF ONLINE RESOURCES AND TOOLS FOR TEACHING STATISTICAL CONCEPTS

These tools are particularly useful in making abstract concepts more tangible and visually appealing. For instance, websites like Khan Academy and Desmos offer interactive modules that cover a wide range of topics in statistics, from basic probability to advanced statistical inference. On platforms like these, students can manipulate data, adjust parameters, and observe how statistical concepts change in response. In the case of probability distributions, for example, students can dynamically adjust parameters and observe how the distribution changes, providing an intuitive understanding of concepts such as variability, skewness, and normality (Poultsakis S., et al, 2021). These resources allow students, allowing them to revisit content as needed, pause lessons, and take notes without the pressure of a traditional classroom setting. For example, Coursera offers courses in partnership with prestigious institutions like Stanford and Harvard, providing students with learning leading the Many these courses include video lectures, practice exercises, and quizzes understanding track their progress (Demszky D., et al, 2024). Moreover, these datasets can be used to teach data-wrangling techniques, such as handling missing data, outliers, and skewed distributions, which are essential skills for working with real data (Pratiwi D I., et al, 2023). LMS platforms make it easy for students to stay organized, access learning materials, and participate in collaborative activities, all of which contribute to a more cohesive and interactive learning experience. Despite the many advantages of using online resources and tools, several challenges need to be addressed to maximize their effectiveness in teaching statistical concepts. The primary concern is to take full advantage of online resources. Instructors should be mindful of this issue and ensure that all students have access to the materials they need, whether through alternative formats or by providing support for those with limited access to technology. Additionally, technical difficulties such as software incompatibility, platform malfunctions, and issues with data access can sometimes disrupt the learning process. To mitigate these challenges, instructors should provide clear instructions on how to use the tools, offer technical support, and troubleshoot common problems (Lapitan Jr L D S., et al, 2021).

REPRESENTATIONS IN DEVELOPING STATISTICAL THINKING: SUMMARY OF KEY FINDINGS AND IMPLICATIONS.

The development of statistical thinking is an essential goal in education, and how statistical concepts are represented plays a crucial role in fostering this type of thinking. Representations, including symbolic, graphical, and verbal formats, serve as tools to support the learning process by helping students conceptualize, process, and apply statistical ideas. These representations not only help learners understand abstract statistical principles but also encourage them to make connections between concepts, visualize data, and interpret results in meaningful ways. Various forms of representation—symbolic, visual, and verbal—are integral in developing students' understanding of statistics, and each type contributes uniquely to the development of statistical thinking. Similarly, a student who relies solely on visual representations may be able to identify trends in data but may lack the deeper understanding necessary to apply statistical reasoning or draw valid conclusions (Sharma H., 2021). Visual representations have been shown to enhance students' ability to interpret data, identify patterns, and make inferences, contributing to the development of statistical reasoning (Fries L., et al, 2021). Additionally, interactive visualizations—tools that allow students to manipulate data, adjust variables, or explore different scenarios—can help a more flexible These tools have the advantage of providing instant feedback, which reinforces learning and allows students to see how changing one aspect of a dataset can affect the overall analysis. However, research has shown that many students struggle to transition from concrete, visual representations to abstract symbolic representations. This difficulty is often referred to as the concreteness fading problem, where students can initially understand statistical concepts when they are presented visually but struggle to work with these concepts when they are represented symbolically. To address this challenge, it is important to support students in making connections between visual and symbolic representations, helping them understand how equations correspond to real-world phenomena and how statistical formulas are derived from underlying patterns in the data (Hawes Z C K., et al, 2022). In addition to visual and symbolic representations, verbal representations—such as written or spoken descriptions, explanations, and discussions—play a vital role in developing statistical thinking. The development of statistical thinking also involves strategies of metacognition (Lyle C., et al, 2021). The integration of various representations not only helps students build an intuitive understanding of statistical ideas but also fosters mastering statistics and applying it effectively in real-world contexts.

FUTURE DIRECTIONS FOR RESEARCH ON REPRESENTATIONS IN STATISTICAL EDUCATION: ANALYSIS OF FUTURE DIRECTIONS FOR RESEARCH ON REPRESENTATIONS IN STATISTICAL EDUCATION

Representations in statistical education play a critical statistical concept, enabling them to engage with abstract ideas in tangible and accessible ways. As statistical literacy becomes increasingly essential in a data-driven society, research into how representations are used, understood, and improved in educational contexts is vital. Several key directions can guide future research in this domain, each aimed at addressing gaps in our understanding and improving teaching practices. The Role of Technology in Representations: Digital tools and technologies have transformed how statistical representations are created, interacted with, and understood. Future research should explore how technology-based visualizations, such as dynamic graphs, interactive simulations, and augmented reality tools, impact students' comprehension of statistical concepts (Rahate A., et al, 2022). This could involve designing representations that are culturally relevant and accessible to learners with visual, auditory, or cognitive impairments. Moreover, studies could examine how language barriers impact students' interpretation of statistical representations and identify strategies to make these tools more universally comprehensible (Garcia J., et al, 2022). Cognitive Processes and Misconceptions: Understanding the cognitive processes underlying students' interactions with statistical representations is another promising research avenue. For example, studies could use eye-tracking technology or think-aloud protocols to investigate how students interpret visualizations and identify common areas of difficulty. Research should also focus on the misconceptions that arise when students misinterpret representations, such as confusing correlation with causation or misreading scales on graphs. By pinpointing the sources of these errors, educators can develop targeted interventions to address them.

Teachers play a pivotal role in guiding students' interactions with statistical representations. Future studies should focus on the preparation and professional development of educators in this area (Sohail S S., et al, 2023). This includes investigating the types of training that help teachers effectively use and explain representations, as well as the challenges they face when implementing these tools in the classroom. Research could also explore how teachers' understanding of statistical representations influences their instructional practices and their students' learning outcomes. Interdisciplinary Approaches: Statistics is inherently interdisciplinary, intersecting fields such as mathematics, science, and social studies. Future research should examine how representations are used across disciplines and how these approaches can inform statistical education. For example, studies could investigate how representations in biology or economics differ from those in traditional statistics courses and whether interdisciplinary approaches enhance students' overall understanding of data analysis. Emerging Areas of Statistical Representation: The rise of big data and advanced statistical techniques, such as machine learning and predictive modeling, presents new challenges and opportunities for representations in statistical education. Future research should explore how to introduce these advanced concepts in ways that are accessible to students. For instance, studies could examine how visualizing machine learning models or large datasets impacts students' understanding and interest in the field. Cross-Cultural Comparisons: Globalization has highlighted the need for statistical literacy across different cultures and educational systems. Comparative studies could investigate how representations are used and interpreted in various cultural and linguistic contexts. This research could identify universal best practices as well as culturally specific strategies for improving statistical education (Aboaoja F A., et al, 2022). Assessment of Representational Competence: Finally, assessing students' ability to use and interpret statistical representations is a critical area for future research. Developing reliable and valid assessment tools that capture the multifaceted nature of representational competence is essential. Research should also explore how assessment results can inform instruction and identify gaps in students' understanding. In conclusion, representations are central to statistical education, serving as both a tool for learning and a window into students' thinking. Future research on representations should be interdisciplinary, inclusive, and technology-driven, addressing both theoretical and practical aspects of statistical teaching and learning. By exploring these directions, educators and researchers can work together to improve the way statistical concepts are taught, ultimately fostering a more statistically literate society.

IMPLICATIONS FOR PRACTICE: USING REPRESENTATIONS TO IMPROVE STATISTICAL EDUCATION: DISCUSSION OF THE IMPLICATIONS FOR PRACTICE, INCLUDING STRATEGIES FOR USING REPRESENTATIONS TO IMPROVE STATISTICAL EDUCATION

Educators should consider principles of visual hierarchy and simplicity, ensuring that students focus on the key message of the representation. For example, color coding, labeling, and consistent use of scales can guide students' interpretation of graphs and charts (Schölkopf B., et al, 2021). Instructional designers and teachers should also account for potential misinterpretations by pre-testing representations with students and refining them based on feedback. Encouraging Representational Fluency: Representational fluency, or the ability to understand and transition between multiple forms of representations, is a cornerstone of statistical literacy. Educators can foster this skill by designing activities that require students to interpret a dataset using different representations, such as creating a bar chart, scatterplot, and box plot from the same data. These exercises help students see how different representations provide unique insights and prepare them to choose the most appropriate form for a given context. Additionally, teachers can scaffold these activities, beginning with simple representations and gradually introducing more complex ones, such as dynamic or multi-dimensional visualizations. Promoting Critical Thinking: Statistical representations are not just tools for understanding data—they are also platforms for critical thinking. Educators should encourage students to question the validity and reliability of representations, analyze potential biases, and explore how visual elements can influence interpretation. For example, activities that involve critiquing real-world graphs or identifying misleading representations in media can cultivate students' critical literacy. By developing these analytical skills, the pervasive use of statistics in modern society. Leveraging Technology: Technology has expanded the possibilities for using representations in statistical education. Tools such as dynamic graphing software, simulations, and interactive dashboards allow students to manipulate data and observe real-time changes in visualizations (Ericsson L., et al, 2022). These interactive experiences can deepen understanding by making abstract concepts tangible. For instance, students can use software to explore how sample size affects a histogram's shape or how outliers influence the mean and median. Teachers should be trained to integrate these tools effectively, ensuring that technology enhances rather than overwhelms learning. Incorporating Real-World Data: The use of real-world data in statistical representations can make learning more relevant and engaging. Teachers should incorporate datasets that reflect students' interests or address significant societal issues, such as climate change, public health, or social justice. By working with authentic data, students not only practice interpreting representations but also understand the practical applications of statistics. Educators should guide students in navigating the complexities of real-world data, including dealing with missing values, variability, and contextual factors.

Training programs should include hands-on workshops where teachers practice creating and interpreting representations, using technology, and addressing common student misconceptions. Additionally, professional development should emphasize pedagogical strategies for fostering representational fluency and critical thinking (Curth A., et al, 2021). Using Representations for Formative Assessment: Representations can be powerful tools for assessing student understanding. For instance, asking students to construct a graph or explain a visual representation provides insight into their conceptual grasp and analytical skills. Teachers can use these formative assessments to identify misconceptions and provide targeted feedback. For example, if students consistently misinterpret the axes of a graph, educators can focus on clarifying this concept through guided practice. Building Collaborative Learning Environments: Representations can also facilitate collaborative learning. Group activities where students analyze or create representations together encourage discussion, peer teaching, and diverse perspectives. These activities can help students refine their understanding through social interaction and shared problem-solving. Teachers should structure these activities to promote meaningful engagement, such as by assigning roles or providing guiding questions. Long-Term Integration Across Disciplines: Statistical representations should not be confined to mathematics or statistics courses but integrated across disciplines such as science, social studies, and economics. For example, students could analyze population growth trends in biology or economic inequality in social studies, using statistical representations as a lens. This interdisciplinary approach reinforces the relevance of statistics and provides students with multiple opportunities to practice and apply their skills (Ziatdinov R., et al, 2022). Monitoring and Improving Practices: Finally, educators and institutions must continuously evaluate the effectiveness of their approaches to using representations.

CONCLUSION

Representations are a cornerstone of developing statistical thinking in secondary education, serving as bridges between abstract statistical concepts and real-world applications. Their use in teaching facilitates the understanding of complex ideas, fosters critical thinking, and enhances students' ability to interpret and analyze data. As secondary education marks a pivotal stage in students' cognitive development, representations play a crucial role in preparing them to navigate the increasingly data-centric world. Effective representations—whether visual, numerical, symbolic, or contextual—empower students to grasp the fundamental concepts of variability, correlation, probability, and inference. These tools allow learners to visualize data patterns, make meaningful connections between different representations, and communicate their findings effectively. Moreover, the ability to translate between representations, such as moving from raw data to a histogram or from a scatterplot to a line of best fit, is essential for cultivating representational fluency—a skill critical for advanced statistical reasoning. The integration of representations into statistical education also enhances students' engagement by contextualizing abstract concepts within real-world scenarios. Authentic datasets, drawn from topics of personal or societal relevance, can motivate students to explore statistical questions and draw conclusions that are meaningful to them. For instance, analyzing trends in climate data, health outcomes, or social justice indicators provides opportunities for students to see the practical applications of their statistical knowledge while simultaneously developing their critical thinking and data literacy skills. Technology further amplifies the potential of representations in secondary education. Tools such as interactive graphing software, dynamic simulations, and real-time data dashboards allow students to engage with data in innovative ways. These resources not only make statistical learning more interactive and accessible but also equip students with the digital skills necessary for modern statistical practices. However, their effective use requires thoughtful integration into teaching practices and professional development for educators. In addition, representations support equity and inclusivity in statistical education. By designing accessible representations and incorporating culturally relevant examples, educators can address the diverse needs of learners. Representations can also be used to scaffold learning, enabling students with varying levels of prior knowledge to build their understanding progressively. Through thoughtful design and implementation, representations can ensure that statistical education is inclusive and equitable, preparing all students to participate in a data-driven society. For teachers, developing expertise in using representations effectively is critical. Professional development programs must equip educators with the skills to select, design, and implement appropriate representations while addressing common student misconceptions. Furthermore, teachers must be prepared to foster discussions around the limitations and biases of representations, encouraging students to approach data critically and ethically. In conclusion, representations are indispensable in fostering statistical thinking in secondary education. They are tools to interpret and communicate data, enabling them to become informed decision-makers and critical thinkers. As educators continue to innovate and adapt teaching practices, representations will remain central to equipping students with the skills and understanding they need for academic success, professional growth, and active citizenship in an increasingly data-driven world. By prioritizing the thoughtful use of representations, secondary education can play a transformative role in developing statistical literacy and empowering the next generation of learners.

REFERENCE

Aboaoja, F. A., Zainal, A., Ghaleb, F. A., Al-Rimy, B A S., Eisa, T. A. E., & Elnour, A. A. H. (2022). Malware detection issues, challenges, and future directions: A survey. *Applied Sciences (Basel, Switzerland), 12*(17), 8482. DOI: 10.3390/app12178482

Alam, A. (2023). "Improving Learning Outcomes through Predictive Analytics: Enhancing Teaching and Learning with Educational Data Mining." 2023 7th International Conference on Intelligent. DOI: 10.1109/ICICCS56967.2023.10142392

Angelis, D., Sofos, F., & Karakasidis, T. E. (2023). Artificial intelligence in physical sciences: Symbolic regression trends and perspectives. *Archives of Computational Methods in Engineering, 30*(6), 1–15. DOI: 10.1007/s11831-023-09922-z PMID: 37359747

Basar, Z. M., Mansor, A. N., & Jamaludin, K. A. (2021). The Effectiveness and Challenges of Online Learning for Secondary School Students–A Case Study. *Journal of University Education, 2*(1), 1–15.

Beranger, B., Lin, H., & Sisson, S. (2023). New Models for Symbolic Data Analysis. *Advances in Data Analysis and Classification, 17*(1), 1–18.

Boyd, R. L., & Schwartz, H. A. (2021). Natural language analysis and the psychology of verbal behavior: The past, present, and future states of the field. *Journal of Language and Social Psychology, 40*(1), 1–15. DOI: 10.1177/0261927X20967028 PMID: 34413563

Burch, M., Ten Brinke, K. B., Castella, A., Peters, G. K. S., Shteriyanov, V., & Vlasvinkel, R. (2021). Dynamic graph exploration by interactively linked node-link diagrams and matrix visualizations. *Visual Computing for Industry, Biomedicine, and Art, 4*(1), 23. DOI: 10.1186/s42492-021-00088-8 PMID: 34491465

Curth, A., & Van der Schaar, M. (2021). On inductive biases for heterogeneous treatment effect estimation. *Advances in Neural Information Processing Systems.*

Demszky D, Liu J, Hill H C. (2024). Can automated feedback improve teachers' uptake of student ideas? Evidence from a randomized controlled trial in a large-scale online course. Journal of Educational Data Mining and Learning Analytics.

Ericsson, L., Gouk, H., Loy, C. C., & Hospedales, T. M. (2022). Self-supervised representation learning: Introduction, advances, and challenges. *IEEE Signal Processing Magazine, 39*(3), 42–62. DOI: 10.1109/MSP.2021.3134634

Estrella, S., Mendez-Reina, M., Olfos, R., & Aguilera, J. (2022). Early statistics in kindergarten: Analysis of an educator's pedagogical content knowledge in lessons promoting informal inferential reasoning. *International Journal for Lesson and Learning Studies, 11*(2), 1–15. DOI: 10.1108/IJLLS-07-2021-0061

Farahani, A. M., Adibi, P., Ehsani, M. S., & Hutter, H. P. (2023). Automatic Chart Understanding: A Review. *IEEE Transactions on Knowledge and Data Engineering, 35*(1), 261–274.

Ferster B. (2023). Interactive visualization: Insight through inquiry.

Franconeri, S. L., Padilla, L. M., & Shah, P. (2021). The Science of Visual Data Communication: What Works. *Science in Context, 34*(2), 143–155. PMID: 34907835

Friedman, H. H. (2023). Cognitive biases and their influence on critical thinking and scientific reasoning: A practical guide for students and teachers.

Fries, L., Son, J. Y., Givvin, K. B., & Stigler, J. W. (2021). Practicing connections: A framework to guide instructional design for developing understanding in complex domains. *Educational Psychology Review, 33*(2), 739–762. DOI: 10.1007/s10648-020-09561-x

Gal, I., & Geiger, V. (2022). Welcome to the era of vague news: A study of the demands of statistical and mathematical products in the COVID-19 pandemic media. *Educational Studies in Mathematics, 109*(1), 1–15. DOI: 10.1007/s10649-022-10151-7 PMID: 35496813

Garcia, J., Villavicencio, G., Altimiras, F., Crawford, B., Soto, R., Minatogawa, V., Franco, M., Martínez-Muñoz, D., & Yepes, V. (2022). Machine learning techniques applied to construction: A hybrid bibliometric analysis of advances and future directions. *Automation in Construction, 142*, 104532. DOI: 10.1016/j.autcon.2022.104532

Harefa D H, Hulu F. (2024). Mathematics learning strategies that support Pancasila moral education: Practical approaches for teachers. Afore: Jurnal Pendidikan Matematika.

Hawes, Z. C. K., Gilligan-Lee, K. A., & Mix, K. S. (2022). Effects of spatial training on mathematics performance: A meta-analysis. *Developmental Psychology, 58*(1), 112–137. DOI: 10.1037/dev0001281 PMID: 35073120

Hosseini, M., Rasmussen, L. M., & Resnik, D. B. (2024). Using AI to write scholarly publications. *Accountability in Research, 31*(7), 715–723. DOI: 10.1080/08989621.2023.2168535 PMID: 36697395

Kamal, T., & Illiyan, A. (2021). School Teachers' Perception and Challenges Towards Online Teaching During COVID-19 Pandemic in India: An Econometric Analysis. *Asian Association of Open Universities Journal, 16*(1), 1–18. DOI: 10.1108/AAOUJ-10-2021-0122

Kiefer, M., Pielke, L., & Trumpp, N. M. (2022). Differential temporo-spatial pattern of electrical brain activity during the processing of abstract concepts related to mental states and verbal associations. *NeuroImage, 247*, 118746. DOI: 10.1016/j.neuroimage.2022.119036 PMID: 35219860

Kliegr, T., Bahník, Š., & Fürnkranz, J. (2021). A review of possible effects of cognitive biases on interpretation of rule-based machine learning models. *Artificial Intelligence, 300*, 103553. DOI: 10.1016/j.artint.2021.103458

Koga, S. (2022). Characteristics of statistical literacy skills from the perspective of critical thinking. *Teaching Statistics, 44*(1), 1–8. DOI: 10.1111/test.12302

Kronberger G, Burlacu B, Kommenda M, Winkler S M. (2024). "ymbolic Regression.

Kumar, D., Haque, A., Mishra, K., & Islam, F. (2023). *Exploring the transformative role of artificial intelligence and metaverse in education: A comprehensive review*. Basic and Applied Sciences.

Lapitan, L. D. S.Jr, Tiangco, C. E., Sumalinog, D A G., Sabarillo, N. S., & Diaz, J. M. (2021). An effective blended online teaching and learning strategy during the COVID-19 pandemic. *Education for Chemical Engineers, 35*, 116–131. DOI: 10.1016/j.ece.2021.01.012

Laupichler, M. C., Aster, A., Schirch, J., & Raupach, T. (2022). Artificial Intelligence Literacy in Higher and Adult Education: A Scoping Literature Review. *Computers & Education, 187*, 1–13. DOI: 10.1016/j.compedu.2022.104441

Lyle, C., Rowland, M., & Ostrovski, G. (2021). *On the effect of auxiliary tasks on representation dynamics*. Machine Learning and Statistics.

Martín, M., & Valiña, M. D. (2023). Heuristics, biases and the psychology of reasoning: State of the art. *Psychology (Irvine, Calif.), 14*(2), 1–15. DOI: 10.4236/psych.2023.142016

Memon M A, Ramayah T, Cheah J H. (2021). PLS-SEM statistical programs: a review. Journal of Applied Structural Equation Modeling.

Mertler C A, Vannatta R A, LaVenia K N. (2021). Advanced and multivariate statistical methods: Practical application and interpretation.

Mutohhari, F., Sutiman, S., Nurtanto, M., & Kholifah, N. (2021). Difficulties in Implementing 21st Century Skills Competence in Vocational Education Learning. *Journal of Research in Education, 31*(2), 1–13.

Nikou, S., & Aavakare, M. (2021). An Assessment of the Interplay Between Literacy and Digital Technology in Higher Education. *Education and Information Technologies, 26*(6), 6715–6733. DOI: 10.1007/s10639-021-10451-0

Nusantara, T., Rahmatina, D., & Purnomo, H. (2021). The Statistical Creative Framework in Descriptive Statistics Activities. *International Journal of Instruction, 14*(3), 1–15.

Poultsakis, S., Papadakis, S., Kalogiannakis, M., & Psycharis, S. (2021). The management of digital learning objects of natural sciences and digital experiment simulation tools by teachers. *Advances in Mobile and Distance Learning, 1*(2), 58–71. DOI: 10.25082/AMLER.2021.02.002

Pratiwi, D. I., & Waluyo, B. (2023). Autonomous Learning and the Use of Digital Technologies in Online English Classrooms in Higher Education. *Contemporary Educational Technology, 15*(2), ep423. DOI: 10.30935/cedtech/13094

Rahate, A., Walambe, R., Ramanna, S., & Kotecha, K. (2022). Multimodal co-learning: Challenges, applications with datasets, recent advances and future directions. *Information Fusion, 81*, 203–239. DOI: 10.1016/j.inffus.2021.12.003

Ringle, C. M., Sarstedt, M., Sinkovics, N., & Sinkovics, R. R. (2023). A perspective on using partial least squares structural equation modelling in data articles. *Data in Brief, 48*, 109074. DOI: 10.1016/j.dib.2023.109074 PMID: 37066088

Ruiz-Rojas, L. I., Acosta-Vargas, P., De-Moreta-Llovet, J., & Gonzalez-Rodriguez, M. (2023). Empowering education with generative artificial intelligence tools: Approach with an instructional design matrix. *Sustainability (Basel), 15*(15), 11524. DOI: 10.3390/su151511524

Schölkopf, B., Locatello, F., Bauer, S., & Ke, N. R. (2021). Toward causal representation learning. *Proceedings of the IEEE.*

Sharma, H. (2021). Statistical significance or clinical significance? A researcher's dilemma for appropriate interpretation of research results. *Saudi Journal of Anaesthesia, 15*(4), 431–434. DOI: 10.4103/sja.sja_158_21 PMID: 34658732

Sitopu J W, Khairani M, Roza M, Judijanto L. (2024). The Importance of Integrating Mathematical Literacy in the Primary Education Curriculum: A Literature Review. International Journal of Teaching and Learning, 1-18.

Sohail, S. S., Farhat, F., Himeur, Y., Nadeem, M., Madsen, D. Ø., Singh, Y., Atalla, S., & Mansoor, W. (2023). Decoding ChatGPT: A taxonomy of existing research, current challenges, and possible future directions. *Journal of King Saud University. Computer and Information Sciences*, *35*(8), 101675. DOI: 10.1016/j.jksuci.2023.101675

Šrol, J. (2022). Individual differences in epistemically suspect beliefs: The role of analytic thinking and susceptibility to cognitive biases. *Thinking & Reasoning*, *28*(1), 1–25. DOI: 10.1080/13546783.2021.1938220

Steed, C. A. (2025). *Interactive data visualization*. Data Analytics for Intelligent Transportation Systems.

Susilawati, E., Khaira, I., & Pratama, I. (2021). Antecedents to Student Loyalty in Indonesian Higher Education Institutions: The Mediating Role of Technology Innovation. *Kuram ve Uygulamada Egitim Bilimleri*, *21*(4), 1329–1344.

Tong, D. H., Uyen, B. P., & Quoc, N. V. A. (2021). The improvement of 10th students' mathematical communication skills through learning ellipse topics. *Heliyon*, *7*(9), e07967. DOI: 10.1016/j.heliyon.2021.e08282 PMID: 34765795

Ünal, Z. E., Ala, A. M., Kartal, G., Özel, S., & Geary, D. C. (2023). Visual and symbolic representations as components of algebraic reasoning. *Journal of Numerical Cognition*, 1–18.

Van Dijke-Droogers, M., Drijvers, P., & Bakker, A. (2022). Introducing statistical inference: Design of a theoretically and empirically based learning trajectory. *International Journal of Science and Mathematics Education*, *20*(8), 1–18. DOI: 10.1007/s10763-021-10208-8

Xu, H., Berres, A., Liu, Y., Allen-Dumas, M. R., & Sanyal, J. (2022). An Overview of Visualization and Visual Analytics Applications in Water Resources Management. *Environmental Modelling & Software*, *147*, 105261. DOI: 10.1016/j.envsoft.2022.105396

Yau, N. (2024). *Visualize This: The FlowingData Guide to Design*. Visualization, and Statistics. DOI: 10.1002/9781394319817

Yau N. (2024). Visualize this: the FlowingData guide to design, visualization, and statistics.

Ziatdinov, R., & Valles, J. R.Jr. (2022). Synthesis of modeling, visualization, and programming in GeoGebra as an effective approach for teaching and learning STEM topics. *Mathematics*, *10*(3), 398. DOI: 10.3390/math10030398

Zou, C., Li, P., & Jin, L. (2021). Online College English Education in Wuhan Against the COVID-19 Pandemic: Student and Teacher Readiness, Challenges and Implications. *PLoS One*, *16*(3), e0248305. DOI: 10.1371/journal.pone.0258137 PMID: 34597337

Chapter 6
Representations in Developing Thinking Regarding Probabilities in Higher Education

Harleen Kaur
https://orcid.org/0000-0002-4656-5248
Chandigarh Group of Colleges, Jhanjeri, India

P. Selvakumar
https://orcid.org/0000-0002-3650-4548
Department of Science and Humanities, Nehru Institute of Technology, Coimbatore, India

S. Haashni
Vimal Jyothi Convent Matriculation Higher Secondary School, India

Nilesh Anute
https://orcid.org/0000-0001-6599-813X
Sri Balaji University, India

Mohit Sharma
https://orcid.org/0009-0007-2280-8077
Maharshi Dayanand University, Rohtak, India

T. C. Manjunath
https://orcid.org/0000-0003-2545-9160
Rajarajeswari College of Engineering, India

ABSTRACT

This chapter is a comprehensive exploration of how various representations in teaching and learning probability influence students' understanding and engagement with this critical area of mathematics. This chapter delves into the cognitive and pedagogical aspects of probability education, emphasizing the role of visual, symbolic, contextual, and interactive tools in fostering conceptual clarity. Probability, being a foundational component of mathematics and statistics, holds immense significance in academic disciplines, professional practices, and everyday decision-making. However, it is also recognized as a challenging subject for students due to its abstract nature and

DOI: 10.4018/979-8-3693-9934-7.ch006

counterintuitive principles. This chapter addresses these challenges by offering insights into the use of representations that simplify complex ideas and help bridge the gap between theory and application. The chapter is organized into sections that explore various dimensions of probability education, such as empirical studies, and practical classroom strategies.

INTRODUCTION TO REPRESENTATIONS IN PROBABILITY EDUCATION: OVERVIEW OF THE CHAPTER AND ITS SIGNIFICANCE

This chapter is a comprehensive exploration of how various representations in teaching and learning probability influence students' understanding and engagement with this critical area of mathematics. This chapter delves into the cognitive and pedagogical aspects of probability education, emphasizing the role of visual, symbolic, contextual, and interactive tools in fostering conceptual clarity. Probability, being a foundational component of mathematics and statistics, holds immense significance in academic disciplines, professional practices, and everyday decision-making. However, it is also recognized as a challenging subject for students due to its abstract nature and counterintuitive principles. This chapter addresses these challenges by offering insights into the use of representations that simplify complex ideas and help bridge the gap between theory and application. The chapter is organized into sections that explore various dimensions of probability education, such as theoretical frameworks, empirical studies, and practical classroom strategies. It examines how representations like probability trees, Venn diagrams, histograms, simulations, and narrative examples support learning by making abstract concepts tangible. A particular focus is placed on the cognitive processes students use to interpret these representations, such as reasoning, problem-solving, and critical thinking. By analyzing how different representations interact with learners' pre-existing knowledge and misconceptions, the chapter provides evidence-based strategies to enhance pedagogical practices. The significance of this chapter lies in its interdisciplinary approach, combining insights from mathematics education, cognitive psychology, and instructional design. It highlights the potential of representations to promote equity and inclusion in the classroom by catering to diverse learning styles and needs. For educators, it offers practical guidance on selecting and implementing representations effectively, ensuring they align with curricular goals and students' developmental stages. By addressing the challenges of abstractness and counter intuitiveness in probability, the chapter makes a compelling case for the strategic use of representations to cultivate a deeper and more meaningful understanding of this essential subject. It not only contributes to the academic discourse but also has practical implications for improving mathematical literacy and critical thinking skills in an increasingly data-driven world.

THE IMPORTANCE OF PROBABILITY EDUCATION IN HIGHER EDUCATION: DISCUSSION OF THE ROLE OF PROBABILITY EDUCATION IN HIGHER EDUCATION

Probability education holds a pivotal role in higher education, serving as a cornerstone for disciplines ranging from mathematics and statistics to engineering, economics, social sciences, and health sciences. As the foundation for understanding uncertainty, randomness, and risk, probability equips students with essential analytical and decision-making skills that are crucial in a world increasingly reliant on data-driven insights. Its relevance extends beyond academic pursuits, influencing critical thinking, logical reasoning, and the ability to interpret real-world phenomena. This discussion highlights the multifaceted role of probability education in higher education, its significance in various disciplines, and the broader impact on personal and professional development. Interdisciplinary Relevance: Probability is integral to multiple academic domains. In mathematics and statistics, it forms the basis for advanced topics such as inferential statistics, stochastic processes, and data science. In engineering, probability underpins concepts like reliability analysis, signal processing, and risk assessment. In economics and finance, it aids in modeling market behaviors, optimizing investments, and understanding economic risks (Michel-Villarreal R., et al, 2023). In the health sciences, it informs evidence-based decision-making through biostatistics and epidemiological modeling. Social sciences leverage probability to study human behavior and societal trends. This universality underscores the need for robust probability education, ensuring that students in diverse fields can grasp, apply, and communicate probabilistic concepts effectively. Probability provides the mathematical foundation for interpreting statistical models, evaluating predictions, and making data-informed decisions (Rahman M M., 2023). As industries increasingly rely on big data and artificial intelligence, higher education institutions are tasked with equipping students with the probabilistic and statistical skills required to thrive in these fields. Whether analyzing trends, predicting outcomes, or optimizing systems, probability education ensures students can meaningfully engage with data in their professional endeavors.

Despite its importance, teaching probability in higher education presents unique challenges. Students often struggle with the abstract nature of probabilistic concepts, such as randomness, independence, and conditional probability (Sailer M., et al, 2021). Misconceptions about probability, rooted in intuitive reasoning, can further hinder understanding. Higher education institutions must address these challenges through innovative pedagogical approaches. Integrating technology, such as simulations and interactive visualizations, can make abstract concepts more tangible. Providing interdisciplinary examples and real-world applications can help students see the relevance of probability in their fields of study. Collaborative learning, problem-based tasks, and case studies can also enhance engagement and comprehension.

Probability education in higher education also plays a crucial role in advancing research and innovation. From designing experiments to analyzing complex systems, probabilistic reasoning is at the heart of scientific inquiry. Graduate students and researchers rely on probability to test hypotheses, model phenomena, and validate findings across disciplines. Institutions that prioritize probability education contribute to the development of a skilled workforce capable of pushing the boundaries of knowledge and innovation (Rasul T., et al, 2023). Moreover, fostering a deep understanding of probability in academia ensures that future educators are well-prepared to teach this vital subject at all levels of education.

In conclusion, probability education is indispensable in higher education, providing students with the tools to navigate uncertainty, interpret data, and make informed decisions in both academic and professional contexts. Its interdisciplinary relevance, role in cultivating critical thinking, and impact on ethical reasoning underscore its importance in a well-rounded education. As the world becomes increasingly complex and data-driven, the need for robust probability education has never been greater. By prioritizing the teaching and learning of probability, higher education institutions can empower students to succeed in their careers, contribute to societal progress, and address the challenges of a rapidly changing world.

CHALLENGES IN TEACHING AND LEARNING PROBABILITY CONCEPTS: ANALYSIS OF THE CHALLENGES FACED BY STUDENTS AND INSTRUCTORS IN PROBABILITY EDUCATION

Probability education poses significant challenges for both students and instructors due to its abstract nature, counterintuitive principles, and the complexities involved in connecting theory to practical applications. These challenges are compounded by common misconceptions, gaps in foundational mathematical knowledge, and the difficulty of aligning instructional approaches with diverse learning needs. Understanding and addressing these obstacles is essential for fostering effective probability education and improving student outcomes. Abstract Nature of Probability: One of the foremost challenges in probability education is the abstract nature of its concepts. Ideas such as randomness, independence, conditional probability, and probability distributions often lack tangible representations, making them difficult for students to grasp. Unlike arithmetic or algebra, where operations are concrete and directly observable, probability deals with events and outcomes that are theoretical or conceptual. For example, understanding the idea of an infinite sequence of random events, as in coin tossing, can be particularly daunting for students who lack exposure to abstract mathematical thinking.

Many concepts in probability contradict students' intuitive reasoning, leading to confusion and misconceptions (Batanero C., et al, 2024). For instance, the Monty Hall problem, where switching doors increases the chance of winning, or the birthday paradox, where the likelihood of shared birthdays is surprisingly high, challenges common assumptions about randomness and likelihood. Bridging the gap between theoretical probability and real-world applications is another significant challenge. While theoretical probability provides the foundation for understanding concepts, students often struggle to apply these ideas in practical contexts, such as statistical modeling, decision-making, or risk assessment (Koparan T., 2022). For example, interpreting probabilistic results in scientific research or analyzing probabilities in financial markets requires a level of contextual understanding that many students find difficult to achieve without extensive practice and guidance.

Probability education often assumes a solid foundation in prerequisite mathematical skills, such as algebra, calculus, and logical reasoning. However, students who lack these foundational skills may struggle to comprehend advanced topics like probability distributions, expected values, or combinatorics. This gap creates a barrier to learning, particularly in higher education, where courses progress rapidly, leaving little room for remediation.

Tools such as simulations, visualization software, and statistical programming can make abstract concepts more tangible and engaging. However, the effective use of technology requires instructors to be proficient in these tools and for students to have access to appropriate devices and resources. Furthermore, over-reliance on technology can sometimes obscure the underlying mathematical principles, leading to superficial understanding (Hokor E K., et al, 2021). Assessing students' understanding of probability concepts is another challenge for instructors. The use of visual tools, such as probability trees, Venn diagrams, and interactive graphs, helps make abstract concepts more accessible. Incorporating technology, such as dynamic simulations and data analysis software, allows students to explore probabilistic scenarios interactively, fostering deeper conceptual understanding (Cai S., et al, 2023). Professional development and training for instructors are also critical for improving probability education. Workshops, online courses, and collaborative teaching communities can help educators develop effective instructional methods and stay updated on best practices.

VISUAL REPRESENTATIONS OF PROBABILITY: GRAPHS, CHARTS, AND DIAGRAMS: EXPLORATION OF VISUAL REPRESENTATIONS OF PROBABILITY CONCEPTS

Visual representations play a vital role in teaching and learning probability concepts, providing intuitive and accessible ways to understand abstract ideas. Graphs, charts, and diagrams help bridge the gap between theoretical knowledge and practical application, enabling students to visualize relationships, identify patterns, and interpret data effectively. By offering concrete, visual structures, these tools enhance comprehension, facilitate problem-solving, and promote critical thinking. This exploration delves into the various types of visual representations used in probability education, their applications, and the ways they address common learning challenges. Probability often involves abstract concepts, such as randomness, independence, and conditional relationships, which are difficult to grasp through symbolic or verbal explanations alone. Visual representations simplify these ideas by presenting them in a more tangible form. They allow students to see the structure of problems, compare outcomes, and analyze data patterns, fostering deeper conceptual understanding (Hu T., et al, 2021). For example, probability trees illustrate decision paths, while histograms show the frequency distribution of outcomes. These tools not only support learning but also help students develop skills in data visualization and interpretation, which are essential in today's data-driven world.

Probability trees are widely used to represent sequential events and their associated probabilities. These diagrams help students visualize the branching of possible outcomes, calculate joint probabilities, and understand conditional relationships. For instance, a probability tree can illustrate the likelihood of drawing different-colored balls from a bag, making it easier to compute probabilities for complex scenarios involving multiple steps.

Venn diagrams are particularly effective for representing set relationships and probabilities involving unions, intersections, and complements. They provide a visual way to explore concepts like mutually exclusive events, independent events, and overlapping probabilities. By shading specific regions, students can directly see the relationships between events and calculate probabilities based on the diagram.

Histograms and bar charts are commonly used to represent frequency distributions and probability distributions. Histograms show how data is distributed across intervals, providing insight into patterns, trends, and variability. For example, a histogram of dice roll outcomes illustrates the uniform distribution of probabilities for each face. Bar charts, on the other hand, are useful for comparing probabilities of discrete events (Das R., et al, 2023). Pie Charts: Pie charts represent probabilities as proportions of a whole, making them useful for visualizing the relative likelihood of different outcomes.

In research and professional settings, these representations are essential for analyzing data, modeling uncertainty, and making decisions. For example, in finance, probability distributions are visualized to assess risk; in healthcare, probability trees are used to evaluate treatment outcomes.

Visual representations offer numerous benefits in probability education: They make abstract concepts more concrete and easier to understand (Efthymiou A., et al, 2021). Visual tools capture students' attention and promote active learning. By providing clear structures, they help students avoid common errors in calculations and reasoning. Visual representations are widely applicable across disciplines, from science and engineering to social sciences and business. They cater to diverse learning styles, including visual learners, and make complex ideas accessible to a broader audience.

Despite their advantages, visual representations also pose challenges. Misinterpretations of graphs or diagrams can lead to misconceptions. For example, students might incorrectly interpret overlapping areas in a Venn diagram or misread scales on a histogram. Furthermore, creating accurate and meaningful visualizations requires careful attention to design and clarity. Instructors must ensure that visuals are appropriately labeled, scaled, and contextualized to avoid confusion. To maximize the effectiveness of visual representations, educators can adopt the following strategies:

Integrate Technology: Use software tools to create dynamic and interactive visualizations (Weiskopf D., 2022).

Provide Context: Relate visuals to real-world scenarios to enhance relevance and engagement.

SYMBOLIC REPRESENTATIONS OF PROBABILITY: MATHEMATICAL NOTATIONS AND FORMULAS: DISCUSSION OF SYMBOLIC REPRESENTATIONS OF PROBABILITY CONCEPTS

Symbolic representations in probability, encompassing mathematical notations and formulas, serve as the universal language for articulating and analyzing probabilistic concepts. These representations are indispensable for expressing ideas with precision, performing complex calculations, and building models that describe uncertainty and randomness. They provide a framework for understanding relationships between events, computing probabilities, and solving real-world problems. However, the abstract and formal nature of symbolic representations often poses challenges for students, requiring careful instruction to ensure comprehension and application.

Symbolic representations provide a structured and concise way to express complex probabilistic ideas. They enable the formulation of general principles, facilitate calculations, and support the development of advanced models. In academic and professional contexts, these representations are essential for communicating ideas, deriving theoretical results, and applying probability to solve practical problems. For instance, symbolic representations are used to model risks in finance, optimize systems in engineering, and analyze data in health sciences.

The abstractness of symbolic notations can be intimidating for students, especially those with limited mathematical backgrounds. Symbols may appear cryptic without proper explanation and contextualization. Students often confuse related concepts, such as independence and conditional probability, due to the similar symbolic forms. For example, interpreting is a common error that stems from misunderstanding the notation. The cognitive demands of working with multiple symbols, formulas, and operations simultaneously can overwhelm learners, particularly when solving complex problems involving conditional probabilities or random variables (Furlong M., et al, 2022). Without concrete examples or applications, symbolic representations may seem disconnected from real-world scenarios, leading to disengagement or rote learning. Use technology, such as computer simulations, to demonstrate symbolic concepts dynamically (Pospiech G., et al, 2021). Provide ample practice opportunities, including exercises that require interpreting, deriving, and applying symbolic representations. Offer detailed feedback to address misunderstandings and reinforce learning.

VERBAL REPRESENTATIONS OF PROBABILITY: LANGUAGE AND COMMUNICATION: ANALYSIS OF VERBAL REPRESENTATIONS OF PROBABILITY CONCEPTS

Verbal representations of probability involve the use of language to express probabilistic ideas, describe uncertainty, and communicate the likelihood of events. These representations are essential for translating mathematical concepts into everyday language, fostering accessibility and understanding across diverse audiences. Verbal communication enables individuals to reason about probabilities, make informed decisions, and convey complex ideas in intuitive ways. However, the use of language in probability is not without challenges, as ambiguity, misinterpretation, and cultural differences can complicate effective communication. This analysis explores the role of verbal representations in probability, their applications, limitations, and strategies for enhancing their clarity and effectiveness.

Probability concepts often need to be articulated verbally to facilitate understanding, especially for those who may lack formal mathematical training. Verbal representations bridge the gap between theoretical and practical applications, enabling concepts to be conveyed in contexts such as education, decision-making, research, and public communication (Dhami M K., et al, 2021). For example, people frequently use verbal probability statements, such as "It's likely to rain today" or "There's a 50-50 chance of success," to express uncertainty. Educators use verbal explanations to introduce and clarify probabilistic ideas, helping students develop intuitive understanding before delving into formal mathematical notation. In fields like medicine, finance, and public policy, professionals use verbal descriptions to explain probabilistic outcomes to non-specialists, such as patients, clients, or stakeholders. Verbal representations often rely on descriptive terms to express likelihood. Words like unlikely, possible, probable, and certain provide qualitative assessments of probability. For instance: Unlikely may imply a probability below 25%. Probability could suggest a likelihood of 50–75%. Numerical probabilities are frequently articulated verbally for accessibility. For example, "There's a 30% chance of rain tomorrow" conveys a quantitative assessment in verbal form. Half the time, this treatment is effective translates a probability of 0.5 into everyday language. Verbal communication often includes conditional probability, expressed as If...then statements. For example: If it's cloudy, there's a higher chance of rain. Such statements help clarify relationships between events and dependencies. Probabilities are sometimes communicated through comparisons, such as This option is twice as likely as that one, providing a relative understanding of likelihoods (Marge M., et al, 2022).

Different cultures and languages may assign varying meanings to probability terms, complicating cross-cultural communication. Words like possible or rare might have nuanced connotations in different contexts. Verbal representations that lack numerical context can be insufficient for precise decision-making (Cheung R W., et al, 2022). For instance, saying the risk is small without specifying the probability may leave room for misinterpretation. The way verbal probabilities are framed can influence perception. For example, describing an event as 90% likely versus 10% unlikely can lead to different interpretations due to framing effects.

Verbal probability statements should include context to clarify meaning. For example, there's a 30% chance of rain in this region tomorrow afternoon is more informative than "it might rain tomorrow." Pairing verbal descriptions with numerical probabilities enhances precision and reduces ambiguity. For instance, this event is likely, with a probability of 70%. Establishing consistent definitions for probability terms can reduce variability in interpretation. Educational and professional guidelines can provide standardized ranges for terms like unlikely or probable. Supplement verbal representations with visual tools like probability scales or pie charts to reinforce understanding. For example, showing a chart alongside a verbal description of 50% likelihood can clarify meaning. Be mindful of framing effects and strive for neutral language. Instead of emphasizing certainty or uncertainty, provide balanced descriptions that accurately reflect probabilities (Pitt B., et al, 2022). Tailor Communication to the Audience: Adapt verbal probability statements to the audience's background and needs.

DEVELOPING PROBABILISTIC THINKING: FROM INTUITION TO FORMAL REASONING: DISCUSSION OF THE DEVELOPMENT OF PROBABILISTIC THINKING IN STUDENTS

Probabilistic thinking, the ability to reason about uncertainty and make predictions based on probability, is a critical cognitive skill in modern education and decision-making. Developing this mode of thought requires progression from intuitive reasoning, often shaped by everyday experiences, to formal reasoning grounded in mathematical principles. Intuitive probabilistic thinking is based on an individual's innate ability to perceive and assess uncertainty without formal instruction (Groth R E., et al, 2021). At an advanced stage, students integrate their understanding of probability with other areas of mathematics, such as statistics and calculus. They can model complex situations involving uncertainty, analyze data, and apply probabilistic reasoning in diverse contexts, from scientific experiments to social decision-making. Intuitive reasoning is often influenced by cognitive biases, such as the Representativeness Heuristic. Assuming that small samples must resemble the population (Shodiqin A., et al, 2021). Availability Heuristic: Judging probabilities based on how easily examples come to mind. Gambler's Fallacy: Believing that past events influence independent outcomes. These biases can hinder the transition to formal reasoning by reinforcing misconceptions.

Group activities and discussions allow students to articulate their reasoning, compare approaches, and address misconceptions collectively (Ingram J., 2024). Collaborative learning fosters deeper understanding by encouraging diverse perspectives. Emphasize the interpretation and meaning of probabilistic results rather than just computations. For example, explain what a 70% chance of rain means in practical terms and discuss how people might use this information to make decisions. Highlight connections between probability and other fields, such as statistics, biology, and economics, to demonstrate its relevance and application. Real-world case studies, such as analyzing weather patterns or understanding medical risks, can make probability more engaging and meaningful. Teach students about common biases and how they influence reasoning. Activities that illustrate these biases, such as experiments on the gambler's fallacy, can help students recognize and overcome flawed thinking. Provide structured support as students progress from intuitive to formal reasoning. Begin with simple problems and gradually introduce more complex scenarios, ensuring that each step builds on prior knowledge. Online platforms, apps, and software can provide interactive and engaging ways to explore probability concepts. Tools like simulations, dynamic graphs, and adaptive quizzes can make learning more interactive and effective. Developing probabilistic thinking equips students with essential skills for navigating a world of uncertainty (Kazak S., et al, 2021). In an era dominated by data and probabilistic models, the ability to reason about likelihoods and make informed predictions is invaluable. From understanding risks in healthcare to analyzing trends in technology and finance, probabilistic thinking has broad applications. Moreover, fostering this skill promotes critical thinking, logical reasoning, and decision-making abilities, preparing students for academic, professional, and personal success.

COGNITIVE BIASES AND HEURISTICS IN PROBABILITY JUDGMENT: ANALYSIS OF COGNITIVE BIASES AND HEURISTICS THAT INFLUENCE PROBABILITY JUDGMENT

Cognitive biases and heuristics play a significant role in shaping how individuals perceive and evaluate probabilities. While heuristics are mental shortcuts that allow people to make judgments quickly and efficiently, they often lead to systematic errors or biases when dealing with complex probabilistic scenarios. These biases can distort probability judgment, resulting in flawed reasoning, suboptimal decisions, and persistent misconceptions. (Cascão A., et al, 2023). Impact: The representativeness heuristic often leads to the conjunction fallacy, where individuals assume that specific conditions are more probable than a general one (e.g., judging a feminist bank teller as more likely than a bank teller).

Availability Heuristic
Definition: Probability is judged based on how easily examples come to mind.
Example: In a medical test scenario, people may overestimate the likelihood of having a disease after a positive test result without considering the disease's low prevalence in the population.
Impact: Base rate neglect can result in overconfidence and misinterpretation of probabilistic data, particularly in diagnostic settings.
Anchoring Bias

Definition: Probability judgments are influenced by an initial value or anchor, even if it is irrelevant.
Example: When asked whether the probability of rain tomorrow is higher or lower than 40%, people tend to base their estimates around the anchor of 40%, regardless of other information (Berthet V., 2022).
Impact: Anchoring bias can distort probability judgments in decision-making, negotiations, and risk assessments.
Overconfidence Bias:
Definition: People often overestimate the accuracy of their probability judgments or the likelihood of favorable outcomes.
Example: An investor may be overly confident in predicting stock performance, ignoring uncertainty and market volatility.
Impact: Overconfidence bias can lead to unrealistic expectations, excessive risk-taking, and poor decision-making.
Confirmation Bias:
Definition: This bias involves favoring information that supports existing beliefs or hypotheses while disregarding contradictory evidence.
Example: A person might selectively recall instances where their lucky number won in a lottery, reinforcing the belief in its effectiveness.
Impact: Confirmation bias reinforces misconceptions and prevents accurate probability judgment by skewing the evaluation of evidence.
Psychological Foundations of Biases and Heuristics: The biases and heuristics in probability judgment are rooted in cognitive limitations and adaptive strategies:
Cognitive Efficiency: Heuristics allow individuals to make rapid decisions without significant mental effort. However, these shortcuts often sacrifice accuracy for speed.
Pattern Recognition: Humans have an innate tendency to identify patterns, even in random data, leading to biases like the gambler's fallacy and representativeness errors.
Provide Feedback and Practice: Repeated exposure to probabilistic problems and constructive feedback can help individuals identify and correct biased reasoning over time (Martín M., et al, 2023).
Debiasing Techniques: Specific interventions, such as reframing questions or using neutral language, can reduce the impact of biases. For instance, presenting probabilities as frequencies (e.g., 1 in 10 instead of 10%) often leads to more accurate judgments.

TEACHING STRATEGIES FOR PROMOTING DEEP UNDERSTANDING OF PROBABILITY CONCEPTS: EXPLORATION OF TEACHING STRATEGIES FOR PROMOTING DEEP UNDERSTANDING OF PROBABILITY CONCEPTS

Teaching probability effectively requires more than just presenting formulas and procedures; it demands fostering a deep understanding of the underlying concepts and their real-world applications. Probability concepts, often abstract and counterintuitive, can be challenging for students to grasp, especially when they bring preconceived notions or cognitive biases to the learning process. Educators must employ innovative and evidence-based teaching strategies to engage students, address misconceptions, and promote meaningful learning. This discussion explores various teaching strategies that can enhance students' comprehension of probability, including active learning, the use of technology, real-world applications, and scaffolding techniques, while addressing common challenges and their solutions (Ghahramani S., 2024).

Probability involves abstract concepts such as randomness, independence, and conditional probability, which are difficult for many students to visualize or relate to real-world scenarios. Students often bring intuitive but flawed reasoning to probability, such as the gambler's fallacy or misunderstandings of randomness. These misconceptions can hinder the learning process if not explicitly addressed. Traditional teaching methods often focus on rote memorization of formulas and procedures without fostering a deeper understanding of why these methods work or how they relate to real-world problems. Without engaging teaching methods, students may perceive probability as boring or irrelevant, leading to low motivation and surface-level learning.

Start by exploring students' intuitive ideas about probability, even if they are incorrect. For example, discuss their reasoning about everyday probabilistic situations, such as lottery odds or weather predictions. Use evidence and examples to challenge flawed reasoning. For instance, use coin-flipping experiments to illustrate that previous outcomes do not influence independent events. Engage students in experiments that demonstrate probabilistic principles. For example: Flip coins or roll dice to explore outcomes and probabilities (Halaweh M., 2023). Conduct group activities where students calculate probabilities based on real-world scenarios, such as sports statistics or game strategies. Encourage group work to allow students to discuss and reason through probability problems collaboratively, promoting deeper understanding through peer interaction.

Medical testing probabilities. Risk assessment in insurance or finance. Cross-Disciplinary Integration: Show how probability applies in diverse fields, such as biology (genetics), engineering (reliability testing), and social sciences (survey analysis). Start with simple, concrete problems and gradually introduce more complex, abstract scenarios. Offer guided practice, feedback, and step-by-step examples to build confidence and understanding before transitioning to independent problem-solving.

Discuss common biases, such as the representativeness heuristic or base rate neglect, and their impact on probability judgment (Kumar D., et al, 2023). This real-world application makes the concept more meaningful and relatable. Students use simulation software to predict weather patterns based on historical data, learning about concepts like probability distributions and confidence intervals in the process (Kilag O K T., et al, 2024). Interactive and contextualized learning fosters interest and motivation.

USING SIMULATION SOFTWARE TO REPRESENT PROBABILITY CONCEPTS: DISCUSSION OF THE USE OF SIMULATION SOFTWARE TO REPRESENT PROBABILITY CONCEPTS.

Simulation software has revolutionized the way probability concepts are taught and understood, offering dynamic and interactive platforms for students and educators to explore abstract ideas in a concrete and engaging manner. By creating virtual environments that mimic real-world scenarios, these tools enable learners to observe, analyze, and experiment with probabilistic phenomena in ways that traditional methods cannot achieve. This discussion explores the benefits, applications, and challenges of using simulation software to represent probability concepts, highlighting its role in enhancing comprehension, engagement, and critical thinking skills.

Visualizing Abstract Concepts: Probability concepts like randomness, independence, and distributions are inherently abstract. Simulation software provides visual and dynamic representations, making these ideas more accessible.
Example: A simulation can graphically depict the Central Limit Theorem by repeatedly sampling data and displaying the resulting distribution (Nelson B., 2021).
Interactive Learning: Simulations encourage active participation, allowing students to manipulate variables, run experiments, and observe outcomes in real-time.
Example: Students can adjust the number of coin flips in a simulation to see how the law of large numbers operates.
Real-Time Feedback: Simulations provide instant results and visualizations, enabling learners to see the consequences of their actions or decisions.

Example: Changing parameters in a probability distribution simulation immediately reflects in the output, helping students understand relationships between variables.
Safe Exploration of Complex Scenarios: Virtual simulations allow for experimentation with complex or impractical real-world scenarios without risks or resource constraints.
Example: Modeling the spread of diseases using probability simulations to understand contagion rates and intervention impacts.
Customization and Scalability: Many simulation tools are customizable, allowing educators to tailor activities to specific learning objectives. Scalability enables simulations to accommodate individual, group, or classroom-wide learning experiences (Thelen A., et al, 2022). Open-source programming languages for designing highly customizable probability simulations and visualizations.
Example: Monte Carlo simulations to estimate probabilities for complex problems.
Tinker Plots: Aimed at younger learners, this tool facilitates exploratory data analysis and probability experiments (Northrup J M., et al, 2022).
Net Logo: A multi-agent simulation environment for modeling complex systems with probabilistic behaviors.
Desmos: Offers interactive graphing tools to visualize probability distributions and statistical relationships.
Technical Barriers: Access to simulation tools may be limited by hardware, software, or internet availability.
Solution: Utilize free, browser-based tools and ensure compatibility with commonly used devices. Use pre-vetted educational tools and design activities with clear goals and instructions.
Interpretation of Results: Students may misinterpret simulation outputs without proper guidance.
Solution: Encourage critical analysis of results and discussions on their implications.

Enhancing the Learning Experience with Simulations: Integrate Simulations into Lesson Plans: Use simulations as supplementary tools to complement traditional teaching methods (van de Schoot R., et al, 2021). For example, pair a lecture on binomial probability with a simulation demonstrating its distribution.

INTERACTIVE VISUALIZATIONS OF PROBABILITY: DYNAMIC GRAPHS AND CHARTS: EXPLORATION OF INTERACTIVE VISUALIZATIONS OF PROBABILITY CONCEPTS

Interactive visualizations are revolutionizing the way probability concepts are taught and understood, providing students with dynamic, engaging, and intuitive tools for exploring statistical principles. By allowing users to manipulate variables, adjust parameters, and observe the results in real-time, these visualizations create a highly interactive learning environment that deepens students' understanding of abstract probabilistic ideas. Dynamic graphs, charts, and other visual tools offer an immersive way to explore randomness, distributions, and statistical relationships, bridging the gap between theoretical knowledge and empirical understanding. Interactive tools allow students to experiment with variables and observe how changes impact outcomes (Ferster B., 2023). This real-time feedback encourages inquiry-based learning, where students actively discover and test hypotheses. Example: A student can adjust the number of sides on a die in a simulation to explore how different sample spaces affect probabilities, or manipulate parameters in a probability distribution to see the shape of the distribution change. When exploring conditional probability, students can adjust the conditions and see how the probability changes in real time, helping them grasp how different conditions affect the likelihood of events. Interactive visualizations encourage exploration by allowing students to test different scenarios and hypotheses (Burch M., et al, 2021).

Simulating random events, generating probability distributions, and analyzing large data sets to study probabilistic patterns.

R and Python (with Matplotlib, Plotly, or Seaborn): R and Python are powerful programming languages for data analysis and visualization. With libraries such as Matplotlib, Plotly, and Seaborn, students can create highly customized and interactive probability visualizations.
Applications: Developing Monte Carlo simulations, exploring different probability distributions, and visualizing large datasets through interactive graphs and charts (Weiskopf D., 2022).
Wolfram Mathematica: Mathematica is a comprehensive computational software used for solving complex mathematical problems and creating interactive visualizations. It offers advanced capabilities for exploring probability theory and generating dynamic simulations.
Applications: Visualizing complex probability distributions, modeling random processes, and conducting statistical analyses through interactive visualizations.
Randomness and Law of Large Numbers:

Purpose: To help students understand how randomness behaves over repeated trials.
Example: An interactive simulation where students flip a coin or roll a die thousands of times to observe the convergence of empirical probabilities to theoretical values, demonstrating the law of large numbers.
Probability Distributions:
Purpose: To visualize the behavior of different probability distributions, both discrete and continuous.
Example: A dynamic graph showing a normal distribution curve that updates as students change the mean and standard deviation, illustrating how these parameters affect the shape and spread of the distribution.
Conditional Probability and Bayes' Theorem:
Purpose: To illustrate how probability changes when additional information is provided.
Example: An interactive tool where students calculate the probability of an event occurring given certain conditions (e.g., the probability of having a disease given a positive test result) and explore how Bayes' theorem applies to real-world scenarios.
Simulating Probability Experiments:
Purpose: To provide hands-on experience with experimental probability.
Example: A virtual dice-rolling tool that allows students to run simulations of thousands of dice rolls and calculate empirical probabilities, comparing them with theoretical predictions.
Expected Value and Risk Analysis:
Purpose: To visualize decision-making under uncertainty and calculate the expected value of different outcomes.
Example: A simulation of a betting game where students adjust the odds and calculate the expected value of a game of chance, helping them understand risk and reward.
Monte Carlo Simulations:
Purpose: To use random sampling to model complex probabilistic systems.
Example: A Monte Carlo simulation of stock market prices or weather patterns, where students adjust parameters such as volatility or probability and observe the long-term outcomes.

Some visualization tools may be too complex for beginners, requiring time and effort to learn how to use them effectively (Korkut E H., et al, 2023).

ONLINE RESOURCES AND TOOLS FOR TEACHING PROBABILITY CONCEPTS: ANALYSIS OF ONLINE RESOURCES AND TOOLS FOR TEACHING PROBABILITY CONCEPTS

The advent of online resources and tools has transformed the way educators teach and students learn probability concepts. With the growing reliance on digital platforms for education, an increasing number of websites, applications, and online tools have emerged to support the teaching of probability. These tools offer various methods for engaging students, from interactive simulations to video tutorials, dynamic graphs, and statistical software, all designed to provide a deeper understanding of probabilistic concepts. This discussion explores the value and significance of online resources in teaching probability, highlighting their advantages, types of tools, and how they can be effectively integrated into the curriculum to enhance student learning outcomes.

One of the key advantages of online resources is the vast array of materials available to students and educators. These resources can include tutorials, instructional videos, interactive exercises, articles, and pre-recorded lectures, offering diverse learning formats that cater to various learning styles.

Example: Students can watch videos explaining the basics of probability and then immediately practice exercises on the same platform to reinforce the concepts they just learned (Demszky D., et al, 2024). Many online tools provide interactive simulations, quizzes, and activities that allow students to engage directly with probability concepts. Games, quizzes, and interactive exercises often incorporate elements of fun, providing a gamified learning experience that encourages students to keep experimenting and learning (Lavidas K., et al, 2022).

Example: Gamified probability activities, where students compete to solve probability puzzles or simulations, can increase engagement and foster friendly competition among peers.

Interactive simulations allow students to experiment with probability concepts in real-time, providing a hands-on approach to learning. These tools are particularly effective for visualizing random events, probability distributions, and statistical experiments.

A free online graphing calculator that enables students to plot probability distributions, visualize statistical data, and simulate probability experiments.

Wolfram Alpha: A computational engine that can solve probability problems, generate probability distributions, and provide step-by-step solutions, helping students learn and verify their understanding.

Quizzes and Problem Sets: Many online platforms offer quizzes and problem sets specifically designed to test students' knowledge of probability concepts. These resources often come with explanations for correct and incorrect answers, helping students improve their understanding through practice and review (Alenezi M., 2023).

Examples: Brilliant.org: Provides interactive problem-solving exercises focused on probability, allowing students to solve puzzles and receive immediate feedback.

Quizlet: Hosts user-generated flashcards and quizzes on probability topics, which can be customized for different learning levels.

Collaborative Projects: Online resources enable students to collaborate on group projects, where they can use tools like online probability simulations and data analysis platforms to investigate real-world probabilistic problems. These projects can foster teamwork, critical thinking, and problem-solving skills.

Example: Students could work together on a project that involves using probability simulations to model and analyze a real-world scenario, such as predicting the weather or simulating the outcomes of a sports game (Gupta S K., et al, 2021).

REPRESENTATIONS IN DEVELOPING THINKING REGARDING PROBABILITIES: SUMMARY OF KEY FINDINGS AND IMPLICATIONS

Visual representations such as diagrams, graphs, and charts are powerful tools in helping students understand abstract probability concepts. Research has shown that visual representations make it easier for students to grasp complex probabilistic phenomena, especially when dealing with concepts such as distributions, randomness, and conditional probability. By allowing students to manipulate variables and visually observe outcomes, these representations provide immediate feedback that enhances comprehension. Visual tools, such as probability trees, Venn diagrams, and bar graphs, help students internalize the structure of probabilistic problems and recognize patterns, facilitating better decision-making processes when faced with uncertainty. Symbolic representations, including mathematical formulas and notations, are foundational in formalizing probability theory (Brown G D., et al, 2021). Research suggests that students often face difficulties transitioning from intuitive or empirical notions of probability to more formal, mathematical reasoning. However, when appropriately scaffolded, symbolic representations allow students to precisely articulate and calculate probabilities, providing the necessary structure for rigorous problem-solving. Interactive simulations and dynamic models are powerful tools that enable students to explore and manipulate probabilistic concepts in real-time. Research has shown that interactive tools that allow students to experiment with probabilistic models and immediately observe outcomes are particularly effective for enhancing conceptual understanding. These dynamic representations not only help students visualize abstract concepts but also provide opportunities for experiential learning and hands-on experimentation (Kruschke J K., 2021). Key Insight: Interactive simulations help bridge the gap between intuition and formal probability theory by giving students the chance to actively engage with probabilistic concepts.

This could involve starting with simple, concrete examples that use visual aids and familiar contexts before moving on to more formal, symbolic representations. Educators should provide ample practice opportunities, gradually introducing more complex problems as students develop their understanding. Teachers could first engage students with simple probability problems that rely on visual models, such as drawing cards or rolling dice (Skivington K., et al, 2021). As students gain confidence, they can be introduced to symbolic representations and more complex problems that require algebraic manipulation. Educators should, therefore, focus on helping students visualize and conceptualize probability before introducing complex formulas and procedures. Practical Application: Rather than immediately diving into complex probability equations, instructors could start with interactive simulations that allow students to explore probabilistic phenomena and develop an intuitive understanding. Once students have a strong conceptual grasp, they can be introduced to the formal mathematics of probability. Technology plays a crucial role in providing interactive and dynamic representations of probability concepts. Tools like interactive simulations, online games, and graphing calculators can engage students in active learning and provide immediate feedback. These technologies allow students to experiment with different probabilistic models and observe how changes in variables affect outcomes, fostering a deeper understanding of the concepts (Relkin E., et al, 2021). Teachers can use platforms like PhET Interactive Simulations or Desmos to allow students to explore probability problems in a hands-on, interactive way. This engagement can help students move beyond passive learning and actively participate in their learning process.

FUTURE DIRECTIONS FOR RESEARCH ON REPRESENTATIONS IN PROBABILITY EDUCATION: ANALYSIS OF FUTURE DIRECTIONS FOR RESEARCH ON REPRESENTATIONS IN PROBABILITY EDUCATION.

The continued development of educational technology, the increasing complexity of probability content, and the evolving pedagogical frameworks all present new avenues for research. As we look to the future, several key areas of investigation emerge, each with the potential to improve how educators teach and students learn probability. One of the most promising areas for future research is the continued investigation into the integration of multimodal representations in probability education. While individual representations such as visual diagrams, mathematical symbols, or verbal explanations have been explored extensively, research into how these different modes can be integrated effectively is still in its early stages (Sohail S S., et al, 2023). Multimodal representations combine the strengths of different forms of communication and provide multiple entry points for understanding.

Future research could explore how students engage with and benefit from integrated multimodal representations that combine visual simulations, symbolic equations, and contextualized verbal explanations within the same learning experience. This could involve designing hybrid learning environments that encourage students to use different modes of representation simultaneously, with a focus on improving cognitive processing and deepening understanding. (Zhou S., et al, 2024). Research in this area could inform the design of age-appropriate educational interventions that align with student's cognitive abilities, helping educators create more targeted instructional strategies that build on students' existing knowledge and progressively challenge them as they mature.

Contextualized representations, such as problems set in real-world contexts (e.g., weather forecasting, risk management, and healthcare decision-making), have been shown to engage students and enhance their understanding of probability by making abstract concepts more tangible. However, there is limited research on how best to incorporate these contextualized representations into the curriculum and how they influence students' ability to transfer their learning to new situations. Future research could focus on understanding how contextualized representations affect student engagement, problem-solving skills, and the transfer of learning. For example, how do students use probability concepts learned in the classroom to solve real-world problems? What types of contexts (e.g., social, environmental, economic) resonate most with students and enhance their learning outcomes? A better understanding of how contextualized representations influence learning could lead to the creation of more relevant and engaging learning materials that motivate students to see the real-world applications of probability, thus improving both retention and application of concepts (Gao C., et al, 2024). I

Probability is a fundamental concept that spans multiple disciplines, including mathematics, economics, engineering, biology, and the social sciences. Yet, research on probability education has typically focused on mathematics classrooms, with less attention paid to how representations can be used to teach probability across different domains. While research has examined various representations themselves, less attention has been paid to the teachers' role in selecting and applying them effectively in the classroom (Lee G G., et al, 2023). Teachers may have varying degrees of comfort with different forms of representation, and their choices can impact how students engage with and understand probability.

IMPLICATIONS FOR PRACTICE: USING REPRESENTATIONS TO IMPROVE PROBABILITY EDUCATION: DISCUSSION OF THE IMPLICATIONS FOR PRACTICE, INCLUDING STRATEGIES FOR USING REPRESENTATIONS TO IMPROVE PROBABILITY EDUCATION.

Given the complexity and often counterintuitive nature of probability, the strategic use of these representations is essential in making probability concepts more understandable, relatable, and easier to apply. This section explores the practical implications of using various representations to improve probability education, highlighting key strategies, their benefits, and potential challenges.

Visual representations, such as graphs, charts, probability trees, and Venn diagrams, provide students with concrete ways to visualize probabilistic phenomena. These representations can make abstract ideas like random events, sample spaces, and conditional probability much more tangible and easier to understand. For instance, probability trees help break down complex probabilistic sequences into simpler, sequential steps, while Venn diagrams visually represent relationships between different sets, such as in the case of overlapping events. Teachers can begin lessons on probability with visual tools that help students conceptualize the basic structure of probability problems. For example, introducing the concept of mutually exclusive events using a Venn diagram, or showing the probability distribution of a die roll with a bar graph, allows students to see the outcomes in a way that abstract formulas cannot immediately provide (Mogavi R H., et al, 2023). These representations are essential for providing students with the formal language needed to describe probabilistic events, calculate probabilities, and engage in more advanced problem-solving. For instance, the notation $P(A \cap B)$ represents the probability of both events A and B occurring simultaneously, and understanding this notation is critical for solving problems in probability theory. Practical Strategy: As students develop their understanding of probability, teachers should gradually introduce symbolic representations after grounding students in visual or contextualized representations. However, language can be both a tool and a source of confusion (Akala B M., 2021).

Interactive representations, which allow students to manipulate variables and observe real-time changes in probabilistic outcomes, are increasingly being used in digital tools and platforms. These interactive models, such as online simulations, virtual games of chance, or dynamic probability calculators, enable students to experiment with different probability scenarios and explore how changes in conditions affect outcomes. The interactivity fosters a sense of agency and encourages students to become active participants in their learning process.

Students often approach probability problems with cognitive biases or misconceptions, such as the gambler's fallacy or the law of small numbers (Liu S., et al, 2022). These biases can lead to flawed reasoning and hinder the development of accurate probabilistic thinking. Representations can play a significant role in addressing these biases by providing visual or symbolic feedback that helps students confront and correct their misconceptions. Assessing students' understanding of probability can be challenging, particularly when it comes to evaluating their ability to translate between different types of representations. To gain a holistic understanding of a student's grasp of the material, educators need to assess not only their ability to calculate probabilities but also their understanding of how different representations are connected and how they can be used to solve problems. Teachers can use assessments that ask students to represent probability problems in multiple ways. For example, students might be asked to solve a problem using a symbolic formula, a visual diagram, and a verbal explanation. This not only tests their ability to perform calculations but also evaluates their deeper understanding of the concepts (Alalwani M H J., et al, 2021). Additionally, formative assessments that provide immediate feedback (such as quizzes or interactive simulations) allow students to identify areas for improvement and refine their understanding. Benefit: This comprehensive approach to assessment provides a richer picture of students' learning and ensures that they are not just memorizing formulas but truly understanding the concepts behind them.

CONCLUSION

The exploration of how representations shape the development of probabilistic thinking in higher education reveals that representations play a pivotal role in transforming students' understanding and engagement with probability. Across multiple domains—from introductory courses to more advanced studies in fields such as statistics, economics, engineering, and the natural sciences—probabilistic thinking is foundational. Students' cognitive development in understanding probability can be significantly enhanced by thoughtfully applying diverse forms of representation, including visual, symbolic, verbal, and interactive forms. The integration of these representations helps bridge the gap between intuitive and formal reasoning, encourages active learning, and fosters a deeper understanding of abstract concepts that are central to probability theory. As students move through their academic careers, the challenge becomes not just to learn probabilistic formulas and calculations, but to cultivate a sophisticated understanding of uncertainty, risk, and variability. The ability to represent probability concepts using multiple modes of communication enables students to engage with these concepts from different angles, which is essential for tackling complex problems in higher education. In the context of higher education, this need for diverse representations is particularly critical given that many of the students who study probability will eventually apply these concepts in multidisciplinary contexts, ranging from decision-making in economics to data analysis in the sciences and social sciences.

One of the key insights from this exploration is the importance of using multiple representations to enhance students' conceptual understanding of probability. Visual tools such as probability trees, Venn diagrams, and histograms allow students to see the abstract nature of probability in more concrete terms. These representations not only help students visualize events and their relationships but also facilitate a more intuitive grasp of concepts like independence, dependence, and conditional probability. The ability to explain probabilistic concepts in one's own words signifies not only understanding but also the ability to teach and transfer knowledge. Furthermore, integrating verbal reasoning with visual and symbolic representations helps students refine their thinking and become more adept at translating between different forms of representation as they progress in their academic studies.

Another significant development in the teaching and learning of probability in higher education is the advent of interactive representations. Digital technologies, including simulation software, data visualization tools, and online interactive platforms, have revolutionized the way probability concepts are taught. Interactive simulations allow students to experiment with probabilities in real time, offering an experiential learning environment where they can manipulate variables, test hypotheses, and visualize outcomes. This active learning approach promotes deeper engagement with probabilistic concepts and encourages students to think critically about the relationships between variables and the inherent uncertainty that governs real-world phenomena.

REFERENCE

Akala, B. M. (2021). Revisiting education reform in Kenya: A case of Competency-Based Curriculum (CBC). *Social Sciences & Humanities Open*, *3*(1), 100129.

Alalwani, M H J, ALhamad A M. (2021). A framework for investigating the effect of unlimited improvement on quality of education. *International Journal of Academic and Applied Research*, *5*(6), 1–10.

Alenezi, M. (2023). Digital learning and digital institution in higher education. *Education Sciences*, *13*(4), 1–10. DOI: 10.3390/educsci13010088

Batanero, C., & Alvarez-Arroyo, R. (2024). Teaching and learning of probability. *ZDM Mathematics Education*, *56*(1), 1–13. DOI: 10.1007/s11858-023-01511-5

Berthet, V. (2022). The impact of cognitive biases on professionals' decision-making: A review of four occupational areas. *Frontiers in Psychology*, *13*, 1024941. DOI: 10.3389/fpsyg.2021.802439 PMID: 35058862

Brown, G. D., Largey, A., & McMullan, C. (2021). The impact of gender on risk perception: Implications for EU member states' national risk assessment processes. *International Journal of Disaster Risk Reduction*, *55*, 1–12. DOI: 10.1016/j.ijdrr.2021.102452

Burch, M., Ten Brinke, K. B., Castella, A., Peters, G. K. S., Shteriyanov, V., & Vlasvinkel, R. (2021). Dynamic graph exploration by interactively linked node-link diagrams and matrix visualizations. *Visual Computing for Industry, Biomedicine, and Art*, *4*(1), 1–15. DOI: 10.1186/s42492-021-00088-8 PMID: 34491465

Burgos, M., Batanero, C., & Godino, J. D. (2021). Algebraization levels in the study of probability. *Mathematics*, *9*(11), 1342. DOI: 10.3390/math10010091

Cai, S., Liu, E., Shen, Y., Liu, C., Li, S., & Shen, Y. (2023). Probability learning in mathematics using augmented reality: Impact on students' learning gains and attitudes. *International Journal of ILEs in Education*, *11*(1), 1–18. DOI: 10.4324/9781003457121-3

Cascão, A., Quelhas, A. P., & Cunha, A. M. (2023). Heuristics and cognitive biases in the housing investment market. *Journal of Housing Markets and Analysis*, *12*(1), 1–18. DOI: 10.1108/IJHMA-05-2022-0073

Cheung, R. W., Hartley, C., & Monaghan, P. (2022). Receptive and expressive language ability differentially support symbolic understanding over time: Picture comprehension in late talking and typically developing children. *Journal of Experimental Child Psychology, 223*, 105564. DOI: 10.1016/j.jecp.2021.105305 PMID: 34653634

Das, R., & Soylu, M. (2023). A key review on graph data science: The power of graphs in scientific studies. *Chemometrics and Intelligent Laboratory Systems, 229*, 104654. DOI: 10.1016/j.chemolab.2023.104896

Demszky, D., Liu, J., Hill, H. C., Jurafsky, D., & Piech, C. (2024). Can automated feedback improve teachers' uptake of student ideas? Evidence from a randomized controlled trial in a large-scale online course. *Educational Evaluation and Policy Analysis, 46*(1), 1–15. DOI: 10.3102/01623737231169270

Dhami, M. K., & Mandel, D. R. (2021). Words or numbers? Communicating probability in intelligence analysis. *The American Psychologist, 76*(1), 1–13. DOI: 10.1037/amp0000637 PMID: 32700939

Efthymiou, A., Rudinac, S., Kackovic, M., & Worring, M. (2021). "Graph neural networks for knowledge enhanced visual representation of paintings." arXiv preprint arXiv:2109.07633. DOI: 10.1145/3474085.3475586

Ferster, B. (2023). *Interactive visualization: Insight through inquiry*. CRC Press.

Furlong, M., & Eliasmith, C. (2022). "Fractional binding in vector symbolic architectures as quasi-probability statements." *Proceedings of the Annual Meeting of the Cognitive Science Society*, 44, 1-13.

Gao, C., Zheng, Y., Wang, W., Feng, F., & He, X. (2024). Causal inference in recommender systems: A survey and future directions. *ACM Transactions on Recommender Systems, 14*(1), 1–35. DOI: 10.1145/3568022

Ghahramani, S. (2024). *Fundamentals of Probability*. Springer.

Groth, R. E., Austin, J. W., Naumann, M., & Rickards, M. (2021). Toward a theoretical structure to characterize early probabilistic thinking. *Mathematics Education Research Journal, 33*(2), 1–20. DOI: 10.1007/s13394-019-00287-w

Gupta, S. K., & Sengupta, N. (2021). Webinar as the future educational tool in higher education of India: A survey-based study. *Technology, Knowledge and Learning, 26*(2), 1–15. DOI: 10.1007/s10758-021-09493-7

Halaweh, M. (2023). ChatGPT in education: Strategies for responsible implementation. *Contemporary Educational Technology, 15*(1), 1–15. DOI: 10.30935/cedtech/13036

Hokor, E. K., & Sedofia, J. (2021). Developing probabilistic reasoning in preservice teachers: Comparing the learner-centered and teacher-centered approaches of teaching. *Journal of Studies in Education and Science, 11*(2), 1–15. DOI: 10.19084/jsses.1114636

Hu, T., Long, C., & Xiao, C. (2021). A novel visual representation on text using diverse conditional GAN for visual recognition. *IEEE Transactions on Image Processing, 30*, 10353–10364. DOI: 10.1109/TIP.2021.3061927 PMID: 33667160

Ingram, J. (2024). Randomness and probability: Exploring student teachers' conceptions. *Mathematical Thinking and Learning, 26*(1), 1–20. DOI: 10.1080/10986065.2021.2016029

Kazak, S., & Pratt, D. (2021). Developing the role of modelling in the teaching and learning of probability. *Research in Mathematics Education, 23*(2), 1–20. DOI: 10.1080/14794802.2020.1802328

Kilag, O. K. T., Malbas, M. H., & Miñoza, J. R. (2024). The views of the faculty on the effectiveness of teacher education programs in developing lifelong learning competence. *Higher Education Studies, 14*(1), 1–12.

Koparan, T. (2022). The impact of a game and simulation-based probability learning environment on the achievement and attitudes of prospective teachers. *International Journal of Mathematical Education in Science and Technology, 53*(4), 531–546. DOI: 10.1080/0020739X.2020.1868592

Korkut, E. H., & Surer, E. (2023). Visualization in virtual reality: A systematic review. *Virtual Reality (Waltham Cross), 27*(1), 1–20. DOI: 10.1007/s10055-023-00753-8

Kruschke, J. K. (2021). Bayesian analysis reporting guidelines. *Nature Human Behaviour, 5*(1), 1–10. PMID: 34400814

Kumar, D., Haque, A., Mishra, K., Islam, F., Kumar Mishra, B., & Ahmad, S. (2023). Exploring the transformative role of artificial intelligence and metaverse in education: A comprehensive review. *Basic and Applied Sciences, 3*(1), 1–15. DOI: 10.56294/mr202355

Lavidas, K., Apostolou, Z., & Papadakis, S. (2022). Challenges and opportunities of mathematics in digital times: Preschool teachers' views. *Education Sciences, 12*(11), 1–12. DOI: 10.3390/educsci12070459

Lee, G. G., Shi, L., Latif, E., Gao, Y., & Bewersdorff, A. (2023). "Multimodality of AI for education: Towards artificial general intelligence." arXiv preprint arXiv:2301.01521.

Liu, S., Liu, S., Liu, Z., Peng, X., & Yang, Z. (2022). Automated detection of emotional and cognitive engagement in MOOC discussions to predict learning achievement. *Computers & Education, 185*, 104442. DOI: 10.1016/j.compedu.2022.104461

Lucena, E. R F C., & Silva, C A T. (2021). The Influence of Cognitive Ability on Cognitive Biases Generated by the Representativeness Heuristic. *Revista Brasileira de Gestão de Negócios, 73*(2), 1–15. DOI: 10.7819/rbgn.v23i1.4090

Mainali, B. (2021). Representation in teaching and learning mathematics. [ERIC.]. *Journal of Education in Mathematics, Science and Technology, 10*(2), 1–12.

Marge, M., Espy Wilson, C., Ward, N. G., Alwan, A., Artzi, Y., Bansal, M., Blankenship, G., Chai, J., Daumé, H.III, Dey, D., Harper, M., Howard, T., Kennington, C., Kruijff-Korbayová, I., Manocha, D., Matuszek, C., Mead, R., Mooney, R., Moore, R. K., & Yu, Z. (2022). Spoken language interaction with robots: Recommendations for future research. *Computer Speech & Language, 74*, 102375. DOI: 10.1016/j.csl.2021.101255

Martín, M., & Valiña, M. D. (2023). Heuristics, biases and the psychology of reasoning: State of the art. *Psychology (Irvine, Calif.), 14*(1), 1–20. DOI: 10.4236/psych.2023.142016

Michel Villarreal, R., & Vilalta-Perdomo, E. (2023). Challenges and opportunities of generative AI for higher education as explained by ChatGPT. *Education, 11*(2), 1–15. DOI: 10.3390/educ11020001

Mogavi, R. H., Deng, C., Kim, J. J., Zhou, P., & Kwon, Y. D. (2023). "Exploring user perspectives on ChatGPT: Applications, perceptions, and implications for AI-integrated education." arXiv preprint arXiv:2301.08221.

Nelson, B. (2021). *Foundations and methods of stochastic simulation.* Springer. DOI: 10.1007/978-3-030-86194-0

Northrup, J. M., Vander Wal, E., Bonar, M., Fieberg, J., Laforge, M. P., Leclerc, M., Prokopenko, C. M., & Gerber, B. D. (2022). Conceptual and methodological advances in habitat-selection modeling: Guidelines for ecology and evolution. *Ecological Applications, 32*(1), 1–15. DOI: 10.1002/eap.2470 PMID: 34626518

Pitt, B., Gibson, E., & Piantadosi, S. T. (2022). Exact number concepts are limited to the verbal count range. *Psychological Science, 33*(1), 1–13. DOI: 10.1177/09567976211034502 PMID: 35132893

Pospiech, G., Merzel, A., Zuccarini, G., & Weissman, E. (2021). "The role of mathematics in teaching quantum physics at high school." Physics: From Research to Teaching, Springer, 1-15. DOI: 10.1007/978-3-030-78720-2_4

Rahman, M. M. (2023). Sample size determination for survey research and non-probability sampling techniques: A review and set of recommendations. *Journal of Entrepreneurship, Business and Management, 11*(1), 1–18. DOI: 10.5281/zenodo.7346384

Rasul, T., Nair, S., Kalendra, D., & Robin, M. (2023). The role of ChatGPT in higher education: Benefits, challenges, and future research directions. *Journal of Applied Learning and Teaching, 6*(1), 1–12. DOI: 10.37074/jalt.6.1.10

Relkin, E., de Ruiter, L. E., & Bers, M. U. (2021). Learning to code and the acquisition of computational thinking by young children. *Computers & Education, 172*, 104224. DOI: 10.1016/j.compedu.2021.104222

Sailer, M., Schultz-Pernice, F., & Fischer, F. (2021). Contextual facilitators for learning activities involving technology in higher education: The C-model. *Computers in Human Behavior, 121*, 103023. DOI: 10.1016/j.chb.2021.106794

Shodiqin, A., Sukestiyarno, Y. L., Wardono, W., & Isnarto, I. (2021). Probabilistic Thinking Profile of Mathematics Teacher Candidates in Problem Solving Based on Self-Regulated Learning. *European Journal of Educational Research, 10*(2), 1–15. DOI: 10.12973/eu-jer.10.3.1199

Skivington, K., Matthews, L., Simpson, S. A., Craig, P., Baird, J., Blazeby, J. M., Boyd, K. A., Craig, N., French, D. P., McIntosh, E., Petticrew, M., Rycroft-Malone, J., White, M., & Moore, L. (2021). Framework for the development and evaluation of complex interventions: Gap analysis, workshop and consultation-informed update. *Health Technology Assessment, 25*(27), 1–128. DOI: 10.3310/hta25570 PMID: 34590577

Sohail, S. S., Farhat, F., Himeur, Y., Nadeem, M., Madsen, D. Ø., Singh, Y., Atalla, S., & Mansoor, W. (2023). Decoding ChatGPT: A taxonomy of existing research, current challenges, and possible future directions. *Journal of King Saud University. Computer and Information Sciences, 35*(4), 14723–14734. DOI: 10.1016/j.jksuci.2023.101675

Thelen, A., Zhang, X., Fink, O., Lu, Y., Ghosh, S., Youn, B. D., Todd, M. D., Mahadevan, S., Hu, C., & Hu, Z. (2022). A comprehensive review of digital twin—part 1: Modeling and twinning enabling technologies. *Structural and Multidisciplinary Optimization, 65*(4), 1–20. DOI: 10.1007/s00158-022-03425-4

Van de Schoot, R., Depaoli, S., King, R., Kramer, B., Märtens, K., Tadesse, M. G., Vannucci, M., Gelman, A., Veen, D., Willemsen, J., & Yau, C. (2021). Bayesian statistics and modelling. *Nature Reviews. Methods Primers*, *1*(1), 1–15. DOI: 10.1038/s43586-020-00001-2

Weiskopf, D. (2022). Uncertainty visualization: Concepts, methods, and applications in biological data visualization. *Frontiers in Bioinformatics*, *2*, 934741. DOI: 10.3389/fbinf.2022.793819 PMID: 36304261

Weiskopf, D. (2022). Uncertainty visualization: Concepts, methods, and applications in biological data visualization. *Frontiers in Bioinformatics*, *2*, 1–12. DOI: 10.3389/fbinf.2022.793819 PMID: 36304261

Zhou, S., Xu, H., Zheng, Z., Chen, J., Li, Z., Bu, J., & Wu, J. (2024). A comprehensive survey on deep clustering: Taxonomy, challenges, and future directions. *ACM Computing Surveys*, *56*(2), 1–37.

Chapter 7
Students' Misconceptions in Learning Probabilities Distributions

Sofia Anastasiadou
https://orcid.org/0000-0001-6404-5003
University of Western Macedonia, Greece

ABSTRACT

The Normal distribution, Binomial distribution, and Poisson distribution, along with other probability distributions taught to economics students, can sometimes be easy to understand and other times challenging. This paper presents, analyzes, and discusses the errors and misconceptions related to their learning. A total of 168 students participated in the survey. They were asked to complete an essay containing eight tasks related to these probability distributions. Implicative statistical analysis was used to examine the data, leading to valuable insights into the misconceptions and mistakes students commonly make. Furthermore, the study explores the implications of these findings.

INTRODUCTION

Statistical education is a fundamental aspect of mathematics teaching in primary, secondary, and higher education. It is a core subject in all university curricula across various disciplines, including social sciences, economics, natural sciences, and medical sciences. Statistics form the foundation for making successful business, medical, economic, and administrative decisions. However, understanding statistical concepts seems to be challenging for students, as these concepts are abstract and require specialized knowledge for comprehension. Education seeks quality and

DOI: 10.4018/979-8-3693-9934-7.ch007

excellence (Taraza et al., 2024; Anastasiadou & Anastasiadis, 2019; Anastasiadou & Taraza, 2019a; 2019b), and its evaluation is specifically based on the European Foundation for Quality Management Excellence (Taraza et al., 2023a; 2023b; Anastasiadou & Taraza, 2020).

Statistical education requires an understanding of probability distributions. However, students face a series of challenges and difficulties. These difficulties include a lack of conceptual understanding, reliance on memorization of formulas without deeper insight into their applications, misinterpretation of statistical concepts, and, most importantly, weak mathematical foundations. Additionally, students encounter obstacles in transitioning from theory to practice, as applying theoretical knowledge to real-world data, problems, and environments often proves to be difficult.

The Normal distribution, Binomial distribution, and Poisson distribution, along with other probability distributions taught to economics students, can sometimes be easy to understand or challenging (Chadjipantelis & Anastasiadou, 2010; Anastasiadou, 2010). The challenges stem from the confusion between different types of distributions—discrete distributions such as the Binomial and Poisson, and continuous distributions such as the Normal. Students often struggle to recognize the appropriate distribution and mistakenly use a continuous distribution, like the Normal distribution, for data that actually follow a discrete distribution, such as the Binomial or Poisson.

Difficulty in identifying the appropriate distribution: Students often fail to select the correct distribution for a given problem. For example, they may apply the Normal distribution to data that follows a Binomial or Poisson distribution (Anastasiadou, 2016). Additionally, they struggle to understand key parameters such as mean, standard deviation, and variance within a distribution. Another significant challenge is distinguishing between the concept of probability and the concept of relative frequency when analyzing a dataset. Students also encounter difficulties in understanding combinatorial problems (Anastasiadou & Loukas, 2009). Furthermore, they fail to differentiate between independent and dependent events, and the algebraic concepts of set theory seem to be particularly challenging for them, as well as the notion of mutually exclusive events (Anastasiadou, 2009a).

Students also face difficulties in understanding the Axiomatic Foundation of Probability, Basic Theorems of Probability, Conditional Probability, Multiplicative Rule of Probability, Joint and Marginal Probabilities, Bayes' Theorem, Multiplication Principle, Permutation, and Combination (Chadjipantelis & Anastasiadou, 2010; Loukas et al., 2008).

In modern educational statistical practice and teaching, the theory of representations receives particular emphasis, as representations are considered inherent to statistical concepts (Kaput, 1987; Goldin, 1998; Duval, 2006; Anastasiadou, 2007; Anastasiadou et al., 2008).

There are cases where representations are so closely tied to a mathematical, statistical, or probability theory concept that it becomes difficult to understand the concept without using the specific representation (Anastasiadou, 2009; Anastasiadou, Chadjipadelis, & Kofou, 2013). Moreover, many researchers emphasize the importance of using representations in different forms or different registers (Anastasiadou et al., 2007; Anastasiadou, 2007; 2019a; 2019b), as well as internal representations, attitudes, and perceptions toward understanding these concepts (Anastasiadou, 2005a; 2005b; 2005c).

This paper presents, analyzes, and discusses the errors and misconceptions related to the learning of probability distributions, such as the Normal distribution, Binomial distribution, and Poisson distribution.

Essay. The essay consists of eight tasks related to these probability distributions, with sub-questions focusing on the concepts of the Normal distribution (V1nn, V2nn), Binomial distribution (V3bb, V4bb), and Poisson distribution (V5pp, V6pp). Additionally, two more tasks involve combinations of the Normal distribution, Binomial distribution, and Poisson distribution (V7nb, V8pb).

As an example, we will present a detailed analysis of the first, third, sixth, and seventh exercises.:

V1nn (Normal distribution Problem). The weight of teenagers (in kg) follows a Normal distribution $N(\mu, \sigma^2) = (55, 10^2)$.

Tasks:

i. Find the probability that a teenager's weight exceeds 65 kg.
ii. Find the probability that a teenager weighs less than 50 kg.

V3bb (Binominal distribution Problem): Research showed that 60 percent of university students used social media daily.

Tasks:

i. Estimate the probability that no more than one student uses social media daily.
ii. Estimate the probability that at least one student uses social media daily

V6pp (Poisson distribution Problem): the mean number of people visiting a university web site with a chat box correspondence during an hour is 24.

Tasks:

i. Find the probability that there are no visits during 12 minutes.
ii. Find the probability that there are at least two visits in 30 minutes.

V7nb (Combination of Normal and Binomial Distributions Problem): The height of teenagers (in cm) follows a Normal distribution: $N(\mu, \sigma^2) = (170, 10^2$, where $\mu = 170$ cm (mean) and $\sigma = 10$ cm (standard deviation).

Tasks:

i. Find the probability that a teenager's height is between 160 cm and 180 cm.
ii. If we choose 4 teenagers randomly, find the probability that exactly two of them have a height between 160 cm and 180 cm.

Sample: A total of 168 students participated in the survey. Implicative Statistical Analysis (SIA) and Principal Components Analysis (PCA) were used to examine the data, leading to valuable insights into the misconceptions and mistakes students commonly make.

Statistical Processing of the Results. For the analysis of the research data, Principal Components Analysis (PCA) and Implicative Statistical Analysis (SIA) were applied (Anastasiadou, S. 2019c; Anastasiadou & Karakos, 2011; Fotiadis & Anastasiadou, 2018a; Anastasiadou, 2018; Gras et al., 1997; Anastasiadou et al., 2024).

During the application of factor analysis, axis rotation (rotation) was performed using the Varimax method (maximum variance rotation) to ensure that the extracted factors (components) were linearly uncorrelated and to maximize the variance among the variable loadings (Fotiadis & Anastasiadou, 2018a; Ntotsi & Anastasiadou, 2019a). This process resulted in smaller loadings becoming even smaller, larger loadings becoming larger, and minimizing those with intermediate values. The eigenvalue or characteristic root criterion, also known as Kaiser's criterion (1960) (Eigenvalue >1), was used to determine the number of retained factors (Hair et al., 1995; Ntotsi & Anastasiadou, 2019b).

The following indices of sample adequacy were used: a) The Kaiser-Meyer-Olkin Measure of Sampling Adequacy (KMO) to assess whether the data were suitable for factor analysis. b) Bartlett's test of sphericity, which examines whether the correlation matrix of the variables included in the analysis differs statistically significantly from the identity matrix. This test ensures that factor analysis is meaningful and also excludes the possibility that all correlation coefficients are zero. For the statistical processing of our research data, we used the Statistical Package for Social Sciences (SPSS), version 29. The significance level for all analyses was set at 5% ($p = 0.05$), except for the test of sphericity, where the significance level was set at 1% ($p = 0.01$) (Anastasiadou, 2014; Anastasiadou, 2009b).

Additionally, the Implicative Analysis by Gras et al. (1997) enables the partitioning and classification of the research variables, as well as the identification of implication relationships between variables or variable classes. The relationships that emerge are not causal relationships but rather a quality index that allows us to assert that a response to one question implies a response to another related question (Fotiadis & Anastasiadou, 2018b; Anastasiadou et al., 2010a; 2010b; 2013).

From Gras's Implicative Statistical Analysis (SIA), the similarity dendrogram was utilized. In the similarity dendrogram, the questions (qi) used in the research are grouped based on the similarity of responses provided by the participants (Lerman, 1981; Couturier & Pazmiño, 2015; Couturier, 2008; Gras & Bodin, 2017). SIA is based on rules and, more specifically, on a probabilistic model (Anastasiadou, 2019c). It represents tendencies within a set of properties and generates association rules (Couturier, 2008). The SIA measure is expressed as a probability, referred to as the intensity of involvement (Anastasiadou, 2019c).

The CHIC program (Classification Hierarchique Implicative et Cohesitive) (Bodin et al., 2000) was used for data analysis through Implicative Statistical Analysis (Couturier & Pazmiño, 2015; Couturier, 2008).

In addition, in order to analysis the data with SIA, a library called RCHIC, available for R is used (Coutourier, 2015) derived from https://members.femto-st .fr/raphael-couturier/en/rchic.

RESULTS

The results revealed differences regarding the students' ability to handle probability distributions with flexibility. The following table demonstrates students' success rate connected with all types of tasks. The results indicate that there are differences in handling different types of distribution given. According to table 1 tasks V1nn and V2nn are the easier ones related to normal distribution as they present the greatest success rate correspondingly. The results revealed differences related to the success in answering every type if the given distribution. The following table shows that tasks V1nn and V2nn did not give any trouble to the students since the success rate were 68.5% and 60.1% correspondingly. It is quite evidence that tasks V1nn and V2nn are favored related to normal distribution.

Tasks V7nb and V8pb seemed to be the most difficult ones related to the combination of normal and poison and normal and binomial distribution and also related regarding poison distribution to the estimation of the parameter λ.

Table 1. Success rate

Task	Incorrect	Half correct	Correct
V1nn	23 (13.7%)	30 (17.9%)	115 (68.5%)
V2nn	27 (16.1%)	40 (23.8%)	101 (60.1%)
V3bb	70 (41.7%)	50 (29.8%)	48 (28.6%)
V4bb	48 (28.6%)	48 (28.6%)	72 (42.9%)
V5pp	64 (38.1%)	58 (34.5%)	46 (27.4%)
V6pp	96 (57.1%)	45 (26.8%)	27 (16.1%)
V7nb	110 (65.5%)	44 (26.2%)	14 (8.3%)
V8pb	123 (73.2%)	32 (19%)	13 (7.7%)

Similarity tree: The similarity tree has two subgroups (Figure 1). The similarity of the first subgroup, subgroup A, ((V7nb V8pb) similarite: 0.772445) is quite strong and it refers to with combination between normal, poison and binomial distribution.

The second subgroup, subgroup B, (V4bb ((V1nn V2nn) (V5pp V6pp)) V3bb)) similarite: 0.959449is very strong. The stronger similarity not only within the group B but in the whole similarity tree is between the items V1nn and V2nn ((V1nn V2nn) similarite: 0.999995), which is almost 1.

The second stronger similarity is between the items V5pp and V6pp (V5pp V6pp) similarite: 0.999966) representing those items related to poison distribution.

The third strongest similarity is among the items V1nn V2nn, V5pp and V6pp (((V1nn V2nn) (V5pp V6pp)) similarite: 0.999576) following by the similarity ((((V1nn V2nn) (V5pp V6pp)) V3bb) similarite: 0.998574) of the whole subgroup B.

The two subgroups, A and B is related with a small similarity of an acceptable importance ((V7nb V8pb) (V4bb (((V1nn V2nn) (V5pp V6pp)) V3bb))) similarite: 0.306037).

Figure 1. Similarity tree

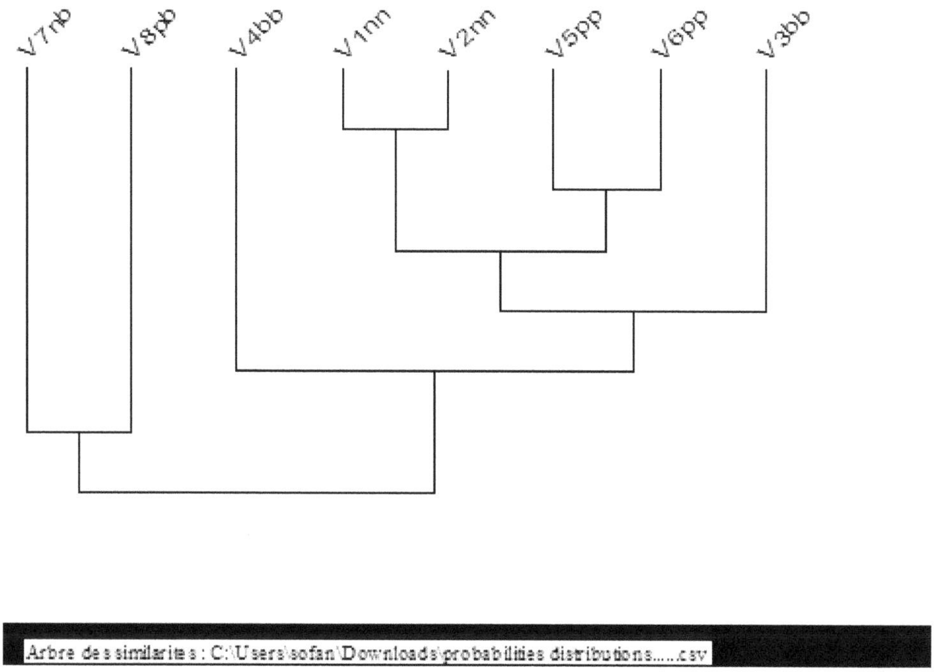

Implicative graph: Implicative graph from a threshold of 0.99 is presented in Figure 2. For example, the rule *V8pb* ⇒ *V7nb* which has the implication intensity 0.99 as it appears in red color. In addition, 58.58% of the participants who gave the right answer to tasks *V8pb* gave the right answer to task *V7nb*.

For example, the rule *V8pb* ⇒ *V6pp* which has the implication intensity 0.99 as it appears in red color. In addition, 87.98% of the participants who gave the right answer to tasks *V8pb* gave the right answer to task *V6pp*.

For example, the rule *V8pb* ⇒ *V5pp* which has the implication intensity 0.99 as it appears in red color. In addition, 73.39% of the participants who gave the right answer to tasks *V8pb* gave the right answer to task *V5pp*.

For example, the rule *V8pb* ⇒ *V3bb* which has the implication intensity 0.99 as it appears in red color. In addition, 67.19% of the participants who gave the right answer to tasks *V8pb* gave the right answer to task *V3bb*.

For example, the rule *V8pb* ⇒ *V2nn* which has the implication intensity 0.99 as it appears in red color. In addition, 72.53% of the participants who gave the right answer to tasks *V8pb* gave the right answer to task *V2nn*.

For example, the rule *V7nb* ⇒ *V3bb* which has the implication intensity 0.99 as it appears in red color. In addition, 72.86% of the participants who gave the right answer to tasks *V7nb* gave the right answer to task *V3bb*.

For instance, the rule $V7nb \Rightarrow V4bb$ which has the implication intensity 0.99 as it appears in red color. In addition, 76.92% of the participants who gave the right answer to tasks *V7nb* gave the right answer to task *V4bb*.

For instance, the rule $V7nb \Rightarrow V2nn$ which has the implication intensity 0.99 as it appears in red color. In addition, 77.92% of the participants who gave the right answer to tasks *V7nb* gave the right answer to task *V2nn*.

For instance, the rule $V7nb \Rightarrow V6pp$ which has the implication intensity 0.99 as it appears in red color. In addition, 88.03% of the participants who gave the right answer to tasks *V7nb* gave the right answer to task *V6pp*.

For instance, the rule $V5pp \Rightarrow V2nn$ which has the implication intensity 0.99 as it appears in red color. In addition, 77.91% of the participants who gave the right answer to tasks *V5pp* gave the right answer to task *V2nn*.

For instance, the rule $V2nn \Rightarrow V1nn$ which has the implication intensity 0.99 as it appears in red color. In addition, 88.27% of the participants who gave the right answer to tasks *V2nn* gave the right answer to task *V1nn*.

Figure 2. Implicative graph

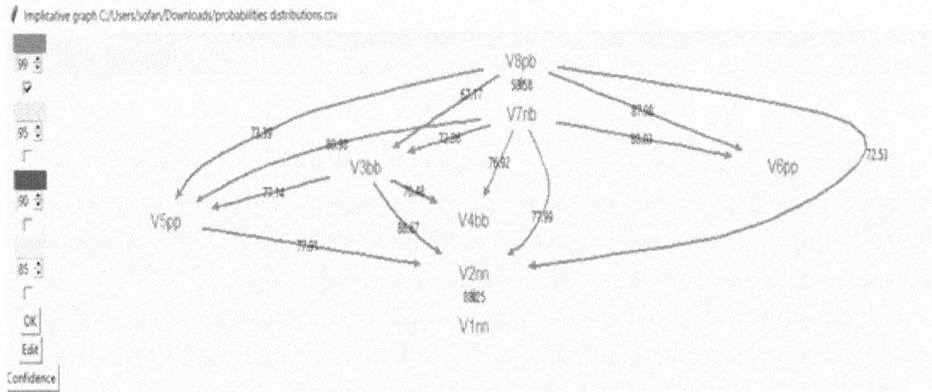

Based on the students' responses and the similarity tree and implication graph it became clear that the easier tasks are related to normal distributions and the combination of continuous and discrete distributions are the most difficult ones.

Principal Components Analysis (PCA) results: Kaiser-Meyer-Olkin Measure of Sampling Ade-quacy well known as KMO and Bartlett's Test of Sphericity Measure of the suitability of Principal Components Analysis application were evaluated before the analysis of the data (Table 2). Both measures were indicators for the permission of the analysis. KMO equals to 0.772 and Bartlett's Test of Sphericity (X^2=807.024, df=28, $p<0.001$) have shown that the application of PCA with varimax rotation is permitted. (Kaiser, 1974).

Table 2. KMO and Bartlett's Test

KMO and Bartlett's Test		
Kaiser-Meyer-Olkin Measure of Sampling Adequacy.		,772
Bartlett's Test of Sphericity	Approx. Chi-Square	807,024
	df	28
	Sig.	<,001

The application of PCA with varimax rotation for all tasks on the basis of eigenvalue criterion in over one was confirmed for three components. These specific components explain 69.133% of the total inertia or variance.

The First Principal Component (PC1), with an eigenvalue equal to 3.346, interprets 29.352% of the total variance of the data, a very satisfactory percentage (Hair, 2005), gathers values for those tasks V2nn and V1nn with very high loadings. These gathered values amount to 0.858 and 0.831 respectively (Table 3). Communalities for tasks V2nn and V1nn count for 0.807 and 0.816 respectively (Table 3) exceeding the 0.40 cut off point criterion, posed as the limit the verification of the satisfactory quality of the tasks. The First Principal Component (PC1) is constructed and interpreted by tasks V2nn and V1nn.

The Second Principal Component (PC2), with an eigenvalue equal to 1.114, interprets 24.102% of the total variance of the data, a very satisfactory percentage (Hair, 2005), gathers values for those tasks V3bb, V5pp, V6pp and V4bb with very high loadings. These gathered values amount to 0.827, 0.745, 0.705 and 0.540 respectively (Table 3). Communalities for tasks V3bb, V5pp, V6pp and V4bb count for 0.766, 0.671, 0.800 and 0.446 respectively (Table 3) exceeding the 0.40 cut off point criterion, posed as the limit the verification of the satisfactory quality of the tasks. The Second Principal Component (PC2) is constructed and interpreted by tasks V3bb, V5pp, V6pp and V4bb.

The Third Principal Component (PC3), with an eigenvalue equal to 1.071, interprets 15.679% of the total variance of the data, a very satisfactory percentage (Hair, 2005), gathers values for those tasks V8pb and V7nb with very high loadings. These gathered values amount to 0.76 and 0.724 respectively (Table 3). Communalities for tasks V8pb and V7nb count for 0.630 and 0.695 respectively (Table 3) exceeding the 0.40 cut off point criterion, posed as the limit the verification of the satisfactory quality of the tasks. The Third Principal Component (PC3) is constructed and interpreted by tasks V8pb and V7nb.

Table 3. Rotated Component Matrix

Rotated Component Matrix[a]				
	Component			
	1	**2**	**3**	**Communalities**
V2nn	.858			.807
V1nn	.831			.816
V3bb		.827		.766
V5pp		.745		.671
V6pp		.705		.800
V4bb		.540		.446
V8pb			.769	.630
V7nb			.724	.695
% of Variance	29.352	24.102	15.679	
Eiguenvalues	3.346	1.114	1.071	
Extraction Method: Principal Component Analysis. Rotation Method: Varimax with Kaiser Normalization				

From the application of Principal Components Analysis (PCA), it becomes evident that students treat the normal distribution in the same way, specifically the two tasks related to normal distribution issues, as reflected in the creation and interpretation of the First Principal Component (PC1).

Students also treat the tasks of discrete distributions in the same way, particularly those related to the Poisson and binomial distributions, as reflected in the creation and interpretation of the Second Principal Component (PC2).

Finally, students handle in a similar manner tasks that require knowledge and problem-solving skills involving a combination of a continuous and two discrete distributions. Specifically, they approach tasks involving the normal distribution combined with the Poisson distribution, as well as the normal distribution combined with the binomial distribution, in the same way, as reflected in the creation and interpretation of the Third Principal Component (PC3).

CONCLUSION-DISCUSSION

This study evaluates the extent to which students can solve problems related to continuous and discrete probability distributions. From the discrete distributions, the Poisson distribution (2 tasks) and the binomial distribution (2 tasks) were selected. From the continuous distributions, two tasks involving normal distribution problems were chosen. Additionally, two tasks combined the normal distribution with the Poisson distribution and the normal distribution with the binomial distribution.

The article presents two different statistical techniques for this purpose, i.e., Principal Components Analysis (PCA) and Implicative Statistical Analysis (ISA), and compares their outcomes based on students' solutions to the given tasks. Furthermore, the two statistical techniques operate complementarily, each highlighting a different dimension in the interpretation of the data.

Students appear to adopt the problem-solving approach for each distribution but struggle to transition smoothly from one type of distribution to another. However, addressing misconceptions in learning probability distributions requires further study in the future, incorporating artificial intelligence (Souravlas & Anastasiadou, 2022; Souravlas et al., 2020), cloud computing (Souravlas et al., 2021a; 2021b; 2021c), as well as cutting-edge technologies and any learning technology and service, taking into account the inevitability of understanding probability distributions (Souravlas et al., 2022a; 2022b; 2023).

The use of interactive tools and simulations, as well as software such as R, Python, and specialized simulation applications, can help students understand concepts through data visualization.

Presenting examples from economics, biology, or social sciences can make learning more engaging and comprehensible, as connecting with real-world data makes the learning process more appealing. Additionally, encouraging students to solve problems based on real-life scenarios can enhance their deeper understanding of distributions.

The use of graphics, analogies, and comparisons between different distributions can help students distinguish the key differences between them, thereby enhancing conceptual understanding.

Understanding probability distributions is a challenge for many students, but with the right educational approach, the learning process can become more effective and meaningful. Incorporating interactive tools, connecting theory with practical applications, and emphasizing conceptual understanding may contribute to improving statistical education and developing stronger data analysis skills.

REFERENCES

Anastasiadou, S. (2002). Enhancing prospective teachers' knowledge of students' attitudes and representations towards statistics-the case of Greek students. In the Journal 'Scientia Paedacogika Experimentals'. *XXXIX*, *1*, 103–127.

Anastasiadou, S. (2004). Teacher's opinions about statistics and probability. In the journal 'Quaderni di Ricerca in Didattica', no 14. pp. 134-143. http://math.unipa .it/~grim/quaderno14.htm

Anastasiadou, S. (2005a). Survey of attitudes and beliefs toward Statistics in Greek Secondary Education International Conference on Mathematics Education, Svishtov, Bulgaria, pp. 125-133.

Anastasiadou, S. (2005b). Affective reactions and attitudes of the last class of Greek high school students towards statistics, In M Boosch (Eds) Proceedings Or the Fourth Conference of the European Society for Research in Mathematics Education (CERME 4), Sant Feliu de Guíxols, Spain, 2005 in the http://cerme4.crm.es/Papers %20definitius/5/Anastasiadou%20Sofia.pdf

Anastasiadou, S. (2005c). Survey of attitudes and beliefs toward Statistics in Greek Secondary Education International Conference on Mathematics Education, Svishtov, Bulgaria, pp. 125-133.

Anastasiadou, S. (2007). Affective Responses and Attitudes of ninth grade Greek Students. *Proceedings of the 5th Mediterranean Conference in Mathematics Education*, Rodes, Greece pp. 169-178.

Anastasiadou, S. (2009a). Greek students' ability in probability problem solving. In the Sixth Conference of the European Society for Research in Mathematics Education. Proceedings of CERME 6, Lyon France © INRP 2010 <www.inrp.fr/ editions/cerme6> pp. 404-412. https://ife.ens-lyon.fr/editions/editions-electroniques/ cerme6/working-group-3

Anastasiadou, S. (2009b). Detection of Factors Influencing Greek Technological Institute Students' Choice of Study Department with the Aid of Correspondence Factor Analysis and Path Analysis. *Journal of the World Universities Forum*, *2*(4), 131–142. DOI: 10.18848/1835-2030/CGP/v02i04/56606

Anastasiadou, S. (2010). Pre-service teachers' performance on the learning of probability distributions and the role of projects: A multilevel statistical analysis. In Regnier, Jean-Claude, Spagnolo, Filippo, Di Paola, Benedetto, & Gras, Regis, Eds, *5th International Conference Implicative Statistic Analysis*, Palermo, Italy pp 73-84.

Anastasiadou, S. (2013). Developing and Evaluating a Structural Equation Model Measuring Leadership Changes in a Lifelong Learning World. *The International Journal of Educational Organization and Leadership*, *19*(2), 1–17. DOI: 10.18848/2329-1656/CGP/v19i02/48520

Anastasiadou, S. (2014). A Structural Equation Model Describes Factors Affecting Greek Students' Consumer behavior. *Procedia Economics and Finance*, *9*(1), 402–406. DOI: 10.1016/S2212-5671(14)00041-0

Anastasiadou, S. (2016). Principal Component Analysis vs. Implicative Statistical Analysis: The case of probability distributions' understanding. Global Journal of Psychology Research. Global Journal on Humanites & Social Sciences. [Online]. 00, pp 00-00. Available from: http://sproc.org/ojs/index.php/pntsbs

Anastasiadou, S. (2018). Comparison of multivariate methods in group/cluster identification. Dissertation thesis of the degree of MSc. in Research Methodology in Biomedicine, Biostatistics and Clinical Bioinformatics, Faculty of Medicine, University of Thessaly.

Anastasiadou, S. (2019a). Pre-service teachers' attitudes toward the use of different modes of representation in statistics teaching and learning. 11th annual International Conference on Education and New Learning Technologies Palma de Mallorca, Spain, EDULEARN 19. pp. 1-5.

Anastasiadou, S. (2019b). Pre-service teachers' attitudes toward the use of different modes of representation in statistics teaching and learning. 11th annual International Conference on Education and New Learning Technologies Palma de Mallorca, Spain, EDULEARN 19. pp. 1-5.

Anastasiadou, S. (2019c). Comparison of contemporary advanced statistical methods regarding construct validity evaluation of TEIque-SF instrument: Statistical Implicative Analysis vs. Principal Components Analysis. *9ème Colloque International sur Analyse Statistique Implicative (ASI 10)*. Belfort – France. pp. 148-163.

Anastasiadou, S., & Anastasiadis, L. (2019). Quality Assurance in Education in the Light of the Effectiveness of Transformational School Leadership. In Sykianakis, N., Polychronidou, P., & Karasavvoglou, A. (Eds.), *Economic and Financial Challenges for Eastern Europe. Springer Proceedings in Business and Economics* (pp. 323–344). Springer., DOI: 10.1007/978-3-030-12169-3_21

Anastasiadou, S., Anastasiadis, L., Angeletos, T., & Vandikas, J. (2010b). A Multidimensional Statistical Analysis of Students' Attitudes toward Physics. *The International Journal of Diversity in Organisations, Communities and Nations*, *16*(5), 341–356. DOI: 10.18848/1447-9532/CGP/v10i01/39794

Anastasiadou, S., Anastasiadis, L., Vandikas, J., & Angeletos, T. (2010a). Implicative statistical analysis and Principal Components Analysis in recording students' attitudes toward electronics and electrical constructions subject. *International Journal of Technology, Knowledge and Society*, *16*(5), 341–356.

Anastasiadou, S., Chadjipadelis, T., & Kofou, I. (2013). The role of projects in helping Greek university students from Pre-school Education department to understand probability distributions International Symposium Elementary Mathematics Teaching (SEMT '13). Prague, Czech Repuplic. International Symposium Elementary Maths Teaching. Proceedings: Tasks and tools in elementary mathematics, Prague [2013] σελ.67-74

Anastasiadou, S., Chadjipadelis, T., & Kofou, I. (2013). Pre-school Education department to understand probability distributions. International Symposium Elementary Mathematics Teaching (SEMT '13), Prague, Czech Republic, In Jarmila Novotná and Hana Moraová ed., pp 67-75.

Anastasiadou, S., & Karakos, A. (2011). The beliefs of electrical and computer engineering students´ regarding computer programming. *International Journal of Technology, Knowledge and Society*, *7*(1), 37–52. DOI: 10.18848/1832-3669/CGP/v07i01/56170

Anastasiadou, S., & Loukas, D. (2009). Greek pre-service teachers' cognitive abilities in understanding the concept of frequency: A multilevel statistical analysis. *International Journal of Learning*, *16*(5), 189–202. DOI: 10.18848/1447-9494/CGP/v16i05/46268

Anastasiadou, S., & Taraza, E. (2019a). Total quality management: implementation of the six sigma methodology for improving quality in higher education, ICERI2019, the 12th annual International Conference of Education, Research and Innovation, Seville (Spain), ICERI2019, pp. 9533-9537.

Anastasiadou, S., & Taraza, E. (2019b). Total Quality Management: Statistical Processes and Quality Organizations' Measurement Tools. ICERI2019, the 12th annual International Conference of Education, Research and Innovation will be held in Seville (Spain), ICERI2019.

Anastasiadou, S., & Taraza, E. (2020). Six Sigma in Tertiary Education: A Win of Change regarding Quality Improvement in Education, Proceedings of of 14th annual International Technology, Education and Development Conference (INTED2020), Valencia, Spain, pp. 9595-9601. DOI: 10.21125/inted.2020.0167

Anastasiadou, S., Zirinoglou, P., Seremeti, L., Masouras, A., Komodromos, M. & Anastasiadis, L. (2024). Statistical Implicative Analysis: A Distinguished Statistical Methodology in Evaluating Greek Students' Attitudes Toward the Use of AI Source Title: Real-World Tools and Scenarios for Entrepreneurship Exploration, (2024), 79-100, .DOI: 10.4018/979-8-3693-3100-2.ch006

Chadjipantelis, T., & Anastasiadou, S. (2010). Pre-service teachers' understanding of probability Distributions: A Multilevel Statistical Analysis. International Conference on the Teaching of Statistics. ICOTS 8, Ljubljanja, Slovenia 2010. https://www.stat.auckland.ac.nz/~iase/publications.php?show=icots8

Couturier, R. (2008). *CHIC: Cohensive Hierarchical Implicative Classification. Studies in Computational Intelligence (SCI).* Springer-Verlag Berlin Heidelberg.

Couturier, R. & Pazmin̄o, R. (2015). Statistical implicative analysis for educational data sets: 2 analysis with RCHI, XVIII Congreso Internacional EDUTEC "Educación y Tecnología desde una visión Transformadora", pp. 1-11

Duval R (2006) A cognitive analysis of problems of com447 prehension in a learning of mathematics. Educ Stud, 448 Math 61:103–131

Fotiadis, Th., & Anastasiadou, S. (2018a). Contemporary advanced statistical methods for the science of marketing: Principal Components Analysis vs Analysee Factorielle des Correspondances. Proceeding of the ICME 2018- 1st International Conference on Marketing and Entrepreneuship. Cyprus. https://hephaestus.nup.ac.cy/handle/11728/11392

Fotiadis, Th., & Anastasiadou, S. (2018b). Contemporary advanced statistical methods for the science of marketing: Implicative Statistical Analysis vs Principal Components Analysis. Proceeding of the ICME 2018- 1st International Conference on Marketing and Entrepreneuship. Cyprus. https://hephaestus.nup.ac.cy/handle/11728/11393

Goldin, G. A. (1998). Representational systems, learning, and 450 problem solving in mathematics. *The Journal of Mathematical Behavior, 451*(17), 137–165. DOI: 10.1016/S0364-0213(99)80056-1

Gras, R., & Bodin, A. (2017). L'A.S.I., Analyseur et révélateur de la complexité cognitive taxonomique. 9ème Colloque International sur Analyse Statistique Implicative, Belfort – France, In Jean-Claude Régnier, Régis Gras, Raphaël Coutourier, Antoine Bodin (edus) pp. 128-142.

Gras, R., Peter, P., Briand, H., & Philippé, J. (1997). Implicative Statistical Analysis. In C. Hayashi, N. Ohsumi, N. Yajima, Y. Tanaka, H. Bock, Y. Baba (Eds.). Proceedingsofthe5th Conference of the International Federation of Classification Societies, Volume 2, pp.412-419. Tokyo, Berlin, Heidelberg, New York: Springer-Verlag.

Hair, J. F., (1995). *Multivariate Data Analysis With Reading* (4th ed.). Prentice Hall.

Kaiser, H. F. (1974). An index of factorial simplicity. *Psychometrika*, *39*(1), 31–36. DOI: 10.1007/BF02291575

Kaput, J. (1987). Representation system and mathematics. In Janvier, C. (Ed.), *Problems of representation in the teaching and learning of mathematics* (pp. 19–26). Lawrence Erlbaum Associates.

Lerman, I. C. (1981). *Classification et Analyse Ordinale des Données*. Dunod.

Loukas, D., Papadimitriou, I., Drosos, G., & Anastasiadou, S. (2008). Investigation of a categorical data distribution via the methods of multivariate data analysis – An application to a stochastic multicriteria group decision-making problem. 67th Meeting of the European Working Group "Multiple Criteria Decision Aiding", Rovaniemi, Lapland, Finland.

Ntotsi, P., & Anastasiadou, S. D. (2019a). Contemporary advanced statistical methods for the science of educational research: Principal components analysis versus L' analysee factorielle des correspondances. *New Trends and Issues Proceedings on Humanities and Social Sciences*. [Online]. 6(1), pp 246–255. Available from from: www.prosoc.eu

Ntotsi, P., & Anastasiadou, S. D. (2019b). Comparison of multivariate pattering methods in group/cluster identification regarding the science of educational research: Implicative Statistical Analysis vs. L' Analysee Factorielle des Correspondances. *New Trends and Issues Proceedings on Humanities and Social Sciences.* [Online]. 6(1), pp 238–245. Available from from: www.prosoc.eu

Souravlas, S., & Anastasiadou, S. (2020). Pipelined Dynamic Scheduling of Big Data Streams. *Applied Sciences (Basel, Switzerland)*, *10*(14), 4796. DOI: 10.3390/app10144796

Souravlas, S., & Anastasiadou, S. (2022). On Implementing Social Community Clouds Based on Markov Models. *IEEE Transactions on Computational Social Systems, 2022*. Advance online publication. DOI: 10.1109/TCSS.2022.3213273

Souravlas, S., Anastasiadou, S., Economides, Th., & Katsavounis, S. 2023. Probabilistic Community Detection in Social Networks. IEEE Access, Journal: IEEE Access, Publication Date: 2023, Volume: 11, On Page(s): 25629-25641, Print ISSN: 2169-3536, Online ISSN: 2169-3536, Digital Object Identifier: DOI: 10.1109/ACCESS.2023.3257021

Souravlas, S., Anastasiadou, S., & Katsavounis, S. (2021). 2021b. A Survey on the Recent Advances of Deep Community Detection. *Applied Sciences (Basel, Switzerland), 11*(16), 7179. DOI: 10.3390/app11167179

Souravlas, S., Anastasiadou, S., & Katsavounis, S. (2021). 2021c. *More on Pipelined Dynamic Scheduling of Big Data Streams.Applied Sciences (Basel, Switzerland), 11*(1), 61. DOI: 10.3390/app11010061

Souravlas, S., Anastasiadou, S., & Kostoglou, I. (2023). 2022b. A Novel Method For General Hierarchical System Modeling Via Colored Petri Nets Based on Transition Extractions From Real Data Sets. *Applied Sciences (Basel, Switzerland), 13*(1), 339. DOI: 10.3390/app13010339

Souravlas, S., Anastasiadou, S., Tantalaki, N., & Katsavounis, S. (2021a). A Dynamic Load Balanced Task Distribution Strategy for Heterogeneous Cloud Platforms Based on Markov Process Modeling. *IEEE Access : Practical Innovations, Open Solutions, 10*, 2169–3536. https://ieeexplore.ieee.org/abstract/document/9729723

Souravlas, S., Anastasiadou, S. D., Tantalaki, N., & Katsavounis, S. 2022a, A Fair, Dynamic Load Balanced Task Distribution Strategy for Heterogeneous Cloud Platforms Based on Markov Process Modeling," in IEEE Access, vol. 10, pp. 26149-26162, 2022, DOI: 10.1109/ACCESS.2022.3157435

Souravlas, S., Katsavounis, S., & Anastasiadou, S. (2020). On Modeling and Simulation of Resource Allocation Policies in Cloud Computing Using Colored Petri Nets. *Applied Sciences (Basel, Switzerland), 10*(16), 5644. DOI: 10.3390/app10165644

Taraza, E., Anastasiadou, S., & Masouras, A. (2023a). Malcolm Baldrige National Quality Award (MBNQA) Quality Tool in Education: A Systematic Literature Review. Proceedings of the European Conference on Innovation and Entrepreneurship, ECIE, 2023, 1, pp. 29–38.

Taraza, E., Anastasiadou, S., Masouras, A., & Papademetriou, C. (2023b). Sustainable Development and Implementation of Quality Management Excellence Models in Public Organizations: A Systematic Literature Review. *Sustainability (Basel), 2023*(15), 7971. DOI: 10.3390/su15107971

Taraza, E., Anastasiadou, S., Papademetriou, C., & Masouras, A. (2024). Evaluation of Quality and Equality in Education Using the European Foundation for Quality Management Excellence Model—A Literature Review. *Sustainability (Basel)*, *2024*(16), 960. DOI: 10.3390/su16030960

Chapter 8
The Role of Multiple Representations in Enhancing Statistical Thinking in Secondary Education:
A Case Study Approach

Vishal Jain
https://orcid.org/0000-0003-1126-7424
Sharda University, India

Archan Mitra
https://orcid.org/0000-0002-1419-3558
NITTE University, India

ABSTRACT

This study explores the role of multiple representations—graphical, numerical, and verbal—in enhancing statistical thinking among secondary school students through a case-based instructional framework. Using a qualitative case study design, data was collected through classroom observations, student artifacts, semi-structured interviews, and reflective journals. The findings reveal that multiple representations significantly improve students' statistical reasoning by supporting diverse cognitive processes, fostering data interpretation, and enhancing problem-solving abilities. However, challenges such as difficulties in transitioning between representations and articulating verbal explanations highlight the need for targeted scaffolding and instructional interventions. This study contributes to educational theory by extending

DOI: 10.4018/979-8-3693-9934-7.ch008

frameworks such as dual coding and cognitive load theories and emphasizes the practical need for teacher training and curriculum development.

1. INTRODUCTION

1.1 Background

Statistical thinking has emerged as a cornerstone of the 21st-century curriculum, driven by the exponential growth of data-driven decision-making across disciplines and professions (Ridgway, 2019). In an era characterized by an unprecedented volume of information, the ability to critically analyze, interpret, and communicate data is indispensable for informed citizenship and professional competence (Pfannkuch & Wild, 2020). Statistical literacy, a critical component of this skill set, empowers individuals to engage with data in meaningful ways, fostering their ability to draw conclusions and make predictions in uncertain situations (Ben-Zvi & Garfield, 2019). Despite its growing importance, embedding statistical thinking effectively in secondary education poses significant challenges, particularly in equipping students with the cognitive and interpretive skills required to engage with complex datasets and representations.

One of the core challenges in developing statistical literacy among secondary school students is the abstract nature of statistical concepts. Unlike arithmetic or algebra, which often involve precise operations with predictable outcomes, statistics frequently require students to grapple with variability, uncertainty, and the interpretation of patterns (Watson, 2021). This abstractness can lead to difficulties in comprehension and application, particularly among students who lack prior exposure to statistical reasoning. Research highlights that learners often struggle to make connections between statistical concepts and real-world contexts, resulting in fragmented understanding and limited transferability of knowledge (Prodromou & Pratt, 2020). Furthermore, traditional pedagogical approaches in many classrooms continue to emphasize rote learning of formulas and procedures, rather than fostering conceptual understanding or critical thinking (Garfield & Ben-Zvi, 2021). Such methods fail to adequately prepare students for the interpretive and communicative demands of modern statistical practices.

The potential of multiple representations as an instructional strategy to overcome these challenges has gained considerable attention in recent years. Multiple representations refer to the use of diverse forms of representation—such as graphs, tables, numerical summaries, verbal explanations, and symbolic notations—to facilitate the understanding of mathematical and statistical concepts (Ainsworth, 2020). These representations serve as cognitive tools that enable learners to approach problems from different perspectives, offering complementary insights into the same phenomena. For example, while a scatterplot provides a visual overview of data trends, a numerical summary highlights specific statistical measures, and a verbal explanation contextualizes the findings within a broader narrative (Chick & Pierce, 2019). By integrating these forms, educators can support students in developing a holistic understanding of statistical concepts, enabling them to move beyond procedural calculations to deeper analytical reasoning.

The theoretical basis for leveraging multiple representations in statistical education is supported by frameworks such as dual coding theory and cognitive load theory. Dual coding theory posits that learners process information more effectively when it is presented through both verbal and visual channels, as this facilitates the creation of richer mental models (Mayer, 2020). Similarly, cognitive load theory emphasizes the importance of reducing extraneous cognitive load by presenting information in accessible and interconnected ways (Sweller et al., 2019). By aligning instructional strategies with these principles, educators can mitigate the cognitive challenges associated with learning statistics, particularly for novice learners who may feel overwhelmed by the complexity of data and concepts.

Empirical studies underscore the efficacy of multiple representations in enhancing statistical thinking. For instance, research by Watson and Donne (2021) demonstrated that secondary students who engaged with statistical problems using a variety of representational forms exhibited greater conceptual understanding and problem-solving flexibility compared to those who relied solely on a single representation. Similarly, Ainsworth (2020) highlighted that tasks incorporating multiple representations encouraged students to make connections between different forms of information, fostering a deeper appreciation of statistical relationships. These findings suggest that the integration of diverse representational forms can not only improve comprehension but also promote critical thinking and metacognitive skills, which are essential for navigating complex statistical tasks.

Despite its potential, the adoption of multiple representations in secondary education remains uneven, with significant variations in implementation across schools and curricula (Ridgway, 2019). Barriers to effective integration include a lack of teacher training, limited availability of instructional resources, and resistance to pedagogical innovation (Pfannkuch & Wild, 2020). Furthermore, the effectiveness of multiple representations is contingent on their thoughtful design and alignment with instructional goals. Poorly designed representations, or those introduced without sufficient scaffolding, can exacerbate cognitive overload and hinder learning (Sweller et al., 2019). Addressing these challenges requires a concerted effort to develop evidence-based instructional strategies that harness the strengths of multiple representations while mitigating their limitations.

The integration of multiple representations in statistical education represents a promising avenue for addressing the challenges faced by secondary students in developing statistical literacy. By leveraging diverse forms of representation, educators can create richer and more engaging learning experiences that promote conceptual understanding, critical thinking, and the ability to apply statistical knowledge in real-world contexts. This study aims to explore the role of multiple representations in enhancing statistical thinking among secondary students, focusing on the cognitive processes and learning outcomes associated with their use in a case-based instructional setting.

Efforts to reduce the maternity penalty and promote gender equity in the workplace have gained significant traction across both national policy landscapes and private-sector initiatives. Case studies from progressive nations and organizations reveal how structural interventions—ranging from equal parental leave to culturally sensitive workplace norms—can help mitigate gender disparities.

Iceland: Equal Parental Leave Model

Iceland stands out globally for its gender-equal parental leave policy, which allocates six months of paid leave for each parent and allows an additional three months to be shared as the family sees fit. A key design feature is its "use-it-or-lose-it" principle for fathers, ensuring that paternity leave cannot be transferred to mothers. Research shows that this model not only increases men's involvement in childcare but also normalizes caregiving roles across genders, thereby reducing the stigma attached to mothers taking extended leave (Eydal & Gíslason, 2014). As a result, Iceland boasts one of the lowest gender pay gaps in the OECD and high maternal employment rates.

Sweden: Incentivizing Shared Paternity

Sweden's parental leave policy offers 480 days of paid leave per child, with 90 days reserved exclusively for each parent. The Swedish government introduced financial incentives, including a gender equality bonus, to encourage fathers to take a more active role. These policies have led to measurable changes in workplace dynamics: organizations are now more accustomed to employees of all genders taking leave, and motherhood no longer automatically implies career disruption. A longitudinal study by Duvander and Johansson (2012) found that women in dual-earner households experienced higher job continuity and faster returns to full-time work when paternity leave was utilized.

Google: Corporate Policy Impact on Retention

In the corporate sector, Google is often cited as a best-practice case in addressing the maternity penalty. In 2007, the company extended paid maternity leave from 12 to 18 weeks, and later to 24 weeks, while improving return-to-work pathways and offering on-site childcare. These changes led to a 50% reduction in attrition among new mothers (Forbes, 2016). Google's strategy demonstrates how aligning parental leave policies with employee needs not only improves gender equity but also benefits companies by reducing hiring and training costs and preserving institutional knowledge.

India: Maternity Benefit (Amendment) Act, 2017

India's Maternity Benefit (Amendment) Act of 2017 extended paid maternity leave from 12 to 26 weeks for women working in establishments with 10 or more employees. It also mandates the option of remote work post-maternity and requires companies with more than 50 employees to provide crèche facilities. While progressive in intent, the policy has had mixed outcomes. Critics argue that the cost burden falls entirely on employers, leading some small and medium enterprises (SMEs) to become hesitant in hiring women of reproductive age (Bhalla & Sanyal, 2018). Nonetheless, the Act represents a crucial step in recognizing maternity-related career disruptions and pushing toward institutional accountability.

1.2 Research Problem

Despite growing recognition of the importance of statistical thinking in education, there remain significant gaps in understanding how multiple representations effectively support its development among secondary school students. Existing research has demonstrated the potential of representations such as graphs, numerical summaries, and verbal explanations to enhance comprehension and analytical reasoning. However, much of this work has been conducted in higher education or adult learning contexts, leaving the unique needs and cognitive challenges faced by younger learners underexplored. Moreover, while theoretical frameworks like dual coding theory and cognitive load theory provide a foundation for the use of multiple representations, empirical studies on their application in real-world secondary education settings are limited. Questions remain regarding how students interact with and integrate these representations in solving statistical problems, as well as the specific ways in which such interactions contribute to deeper learning. This lack of clarity poses a challenge for educators seeking evidence-based strategies to implement multiple representations effectively in their classrooms.

1.3 Research Objective

The primary objective of this research is to investigate the cognitive processes and learning outcomes facilitated by the use of multiple representations in case-based instructional settings. This study aims to explore how diverse forms of representation—such as visual, numerical, and verbal modes—support secondary students in developing statistical thinking. The research will focus on understanding how students transition between these representations, how they use them to solve problems, and how their engagement with multiple forms of representation enhances their conceptual understanding. By examining these interactions within the context of case-based learning, the study seeks to identify practical approaches to integrating multiple representations into secondary education curricula. This objective is rooted in the need to provide a detailed and nuanced understanding of how representations function as cognitive tools, with the goal of improving pedagogical strategies and student outcomes in statistical education.

1.4 Significance of the Study

This study has significant implications for both educational theory and practice. By addressing the research gaps related to the use of multiple representations in secondary statistical education, the findings can contribute to a deeper theoretical understanding of how these tools support cognitive development and learning processes. Such insights can enhance existing frameworks, such as dual coding and cognitive load theories, by providing empirical evidence specific to the context of statistical education.

In practice, the study's findings have the potential to inform curriculum design and teacher training programs. A clearer understanding of the role of multiple representations in facilitating statistical thinking can guide the development of instructional materials that are more aligned with students' cognitive needs. For teachers, this research can provide actionable strategies for designing lessons that effectively incorporate diverse forms of representation, thereby enhancing student engagement and comprehension. Ultimately, the study aims to bridge the gap between theory and practice, offering both a robust conceptual foundation and practical recommendations for improving statistical education in secondary schools.

2. LITERATURE REVIEW

2.1 Overview of Statistical Thinking

Statistical thinking refers to the ability to understand, interpret, and critically analyze data to make informed decisions and predictions in uncertain situations. It encompasses recognizing patterns, understanding variability, and drawing meaningful conclusions from data (Moore et al., 2021). Key components of statistical thinking include the formulation of questions, data collection, data analysis, and the interpretation and communication of findings (Groth, 2019). These skills form the backbone of data literacy, which has become increasingly vital in today's data-centric world.

In secondary education, statistical thinking plays a critical role in fostering analytical skills and preparing students for future academic and professional pursuits (Franklin et al., 2020). Secondary education serves as a formative stage where students develop foundational knowledge of statistical concepts, such as measures of central tendency, variability, and graphical representations of data. However, the teaching of statistics often focuses heavily on procedural aspects, such as memorizing formulas, rather than promoting conceptual understanding or critical thinking (Zieffler & Catalytics, 2020). This traditional approach risks leaving students ill-prepared for real-world challenges, where the ability to contextualize and interpret data is far more important than rote calculations.

Furthermore, the importance of statistical thinking in secondary education is underscored by its interdisciplinary relevance. Statistics is not confined to mathematics but intersects with subjects such as science, social studies, and economics (Watson & Callingham, 2021). Students who develop statistical thinking can better engage with evidence-based reasoning across disciplines, enhancing their overall academic competence and civic awareness. This underscores the need for instructional strategies that go beyond procedural knowledge to develop higher-order thinking skills in statistics.

2.1.1 Gender Disparities in the Workplace: Theoretical and Empirical Foundations

Gender disparities in career progression remain a persistent concern in organizational and policy discourses. Scholars have long explored the structural and psychological factors that contribute to unequal advancement opportunities between men and women. One of the foundational frameworks for understanding this phenomenon is Eagly and Carli's (2007) concept of the "labyrinth of leadership," which posits that women's professional trajectories are not blocked by a single glass ceiling, but rather interrupted by a complex set of cultural, institutional, and interpersonal barriers. These include biased perceptions of competence, the undervaluation of caregiving roles, and stereotypical leadership norms that favor assertive, male-associated traits.

The "motherhood penalty" is a key mechanism through which gender disparities are reinforced. Budig and England (2001) empirically demonstrated that mothers suffer a significant wage penalty compared to childless women and men, even when controlling for work experience and education. The penalty is not merely economic but extends to career stagnation, limited leadership opportunities, and diminished access to high-visibility assignments. The workplace often operates under the "ideal worker norm," which assumes constant availability and uninterrupted labor—a standard that disproportionately disadvantages mothers returning from maternity leave.

The issue becomes more layered when viewed through the lens of intersectionality. Crenshaw (1991) introduced the concept to capture how overlapping identities—such as race, class, and gender—compound the effects of discrimination. For instance, women of color or those from marginalized socioeconomic backgrounds often face even steeper penalties post-maternity, due to compounded structural inequities and cultural stereotypes about their roles as both workers and mothers. Intersectionality thus underscores the need for policy frameworks and workplace reforms that do not treat women as a homogenous category.

Work-life balance literature further illustrates the challenges faced by working mothers. Hochschild's (2003) work on the "second shift" reveals how women, even when employed full-time, continue to bear a disproportionate burden of domestic and caregiving responsibilities. This imbalance exacerbates stress and limits their availability for professional development, networking, or leadership roles. Kossek and Ozeki (1998) reinforce this by highlighting how work-family conflict is a predictor of lower job satisfaction and increased turnover intentions, particularly among mothers without access to flexible work arrangements or reliable childcare.

Cross-cultural and international comparisons shed light on how institutional frameworks can either mitigate or exacerbate the maternity penalty. Scandinavian countries such as Sweden, Norway, and Iceland have introduced **equitable parental leave policies**, state-sponsored childcare, and gender quotas for leadership roles. These policies have contributed to higher female labor force participation and narrower gender pay gaps. In contrast, countries like the United States—where maternity leave is unpaid and often employer-dependent—tend to show larger disparities in women's post-maternity career outcomes. In India, while the Maternity Benefit (Amendment) Act of 2017 mandates 26 weeks of paid leave, informal sector workers and those in smaller firms often remain excluded from its benefits, reinforcing economic insecurity and career disruption.

This comparative lens reveals that while the maternity penalty is a global phenomenon, its magnitude and impact are shaped by cultural norms, legal frameworks, and workplace practices. Therefore, any discussion of gender disparities in career progression must account for both structural barriers and localized contexts, while adopting an intersectional and globally informed approach to policy reform and organizational strategy.

2.2 Multiple Representations in Education

Multiple representations refer to the use of various forms of representation—such as graphs, numerical summaries, verbal descriptions, and symbolic notations—to convey and understand concepts (Kaput, 2021). Each representation offers unique insights into a given problem, and their integration provides a more comprehensive understanding than any single form could achieve. For example, while a bar chart visually highlights patterns in categorical data, a numerical summary quantifies those patterns, and a verbal explanation contextualizes them in real-world terms (Lehrer & Schauble, 2022).

The theoretical foundations of multiple representations are rooted in cognitive psychology. Dual coding theory posits that individuals process information more effectively when it is presented through both verbal and visual modalities, as this facilitates the creation of interconnected mental models (Paivio, 2020). Similarly, cognitive load theory emphasizes the importance of reducing extraneous cognitive load by organizing information in accessible formats that align with learners' cognitive capacities (Sweller, 2021). These theories suggest that multiple representations can support learning by leveraging complementary cognitive resources and reducing the cognitive burden of complex tasks.

Empirical research supports the benefits of multiple representations in education. A study by Goldin and Shteingold (2021) found that students who used multiple representations to solve mathematical problems demonstrated greater flexibility and accuracy than those who relied on a single representation. Similarly, Wilkerson et al. (2022) observed that integrating verbal explanations with visual aids in statistics education improved students' ability to interpret and communicate data. These findings highlight the potential of multiple representations to enhance both comprehension and problem-solving skills.

Despite these benefits, the effective use of multiple representations in education is not without challenges. Poorly designed representations or those introduced without sufficient scaffolding can increase cognitive load and hinder learning (Ainsworth et al., 2020). Additionally, students often struggle to transition between different forms of representation, which may limit their ability to integrate and apply knowledge effectively (Yerushalmy, 2021). Addressing these challenges requires a thoughtful and evidence-based approach to the design and implementation of multiple representations in instructional settings.

2.3 Case-Based Learning

Case-based learning (CBL) is an instructional approach that uses real-world scenarios or cases to engage students in active learning and problem-solving. This approach is particularly effective in fostering deeper understanding and critical thinking, as it situates learning in meaningful contexts (Jonassen & Rohrer-Murphy, 2020). In the context of statistics education, CBL allows students to apply statistical concepts to analyze and interpret real-world data, thereby bridging the gap between theoretical knowledge and practical application (Keeler & Steinhorst, 2022).

One of the key advantages of CBL is its ability to enhance student engagement. By presenting problems that are relevant and authentic, CBL motivates students to explore and connect with the material (Prince & Felder, 2021). For example, analyzing the statistical trends of climate change or public health data can make abstract concepts more tangible and relatable. Additionally, CBL encourages collaborative learning, as students often work in groups to discuss and solve cases, fostering communication and teamwork skills alongside subject-specific knowledge (Herreid & Schiller, 2022).

Research supports the efficacy of CBL in improving learning outcomes in statistics education. A study by Gallagher et al. (2021) found that students engaged in case-based activities demonstrated significantly higher levels of statistical reasoning and problem-solving than those in traditional lecture-based settings. Similarly, Lajoie et al. (2022) observed that CBL improved students' ability to interpret data within broader contexts, enhancing their overall statistical literacy. These findings underscore the potential of CBL as a powerful pedagogical tool for developing statistical thinking.

2.4 Research Gaps and Rationale

Despite the growing interest in multiple representations and case-based learning, several research gaps remain. Most existing studies have focused on higher education or adult learners, with limited attention to secondary education contexts (Lehrer et al., 2021). The cognitive and developmental needs of secondary students differ significantly from those of older learners, necessitating targeted research to understand how these instructional strategies can be adapted to their specific requirements. Additionally, while theoretical frameworks such as dual coding and cognitive load theories provide valuable insights, there is a lack of empirical evidence linking these frameworks to the practical implementation of multiple representations in secondary classrooms (Goldin et al., 2022).

Another significant gap lies in the integration of multiple representations and case-based learning. Although both approaches have been shown to enhance learning outcomes independently, few studies have explored their combined effects on statistical thinking. Understanding how these strategies interact to support cognitive and metacognitive processes could provide valuable insights for curriculum design and instructional practice (Keeler & Steinhorst, 2022). Furthermore, there is a need for longitudinal research to examine the long-term impact of these strategies on students' statistical literacy and reasoning skills (Prince & Felder, 2021).

In light of these gaps, this study aims to contribute to the growing body of research on statistical education by investigating the role of multiple representations in enhancing statistical thinking within a case-based instructional framework. By focusing on secondary education, the study seeks to address the unique challenges and opportunities associated with fostering statistical literacy among younger learners. The findings are expected to provide both theoretical and practical insights, informing the development of more effective and engaging instructional strategies for teaching statistics.

3. RESEARCH METHODOLOGY

3.1 Research Design

This study adopts a qualitative case study approach to investigate how multiple representations support the development of statistical thinking among secondary school students. A case study design is particularly suited to this research, as it allows for an in-depth exploration of complex phenomena within their real-world contexts (Yin, 2020). The study focuses on the interactions between students and various representational forms in a case-based instructional setting, aiming to uncover the cognitive processes and learning outcomes associated with these interactions.

3.2 Participants and Context

The study will be conducted in a secondary school setting, involving students aged 14–16 years enrolled in a statistics-focused mathematics curriculum. A purposive sampling strategy will be employed to select participants who represent a diverse range of abilities, ensuring that findings reflect the experiences of both high-achieving and struggling learners (Creswell & Poth, 2018). A total of 30 students will be selected, divided into smaller groups for case-based activities. Teachers who facilitate these sessions will also be included to provide additional perspectives on instructional strategies and student engagement. This study was conducted in a co-educational, English-medium secondary school located in an urban area of Bangalore, India. The institution follows the Central Board of Secondary Education (CBSE) curriculum, which includes foundational modules in data handling and introductory statistics from Grade 8 onwards. The school serves a socioeconomically diverse student body. A purposive sampling strategy was employed to ensure variation in academic performance, gender, and linguistic background. The sample consisted of 30 students, aged 14–16 years, currently enrolled in Grade 10. The group included 16 female and 14 male students. Based on school records and teacher recommendations, students were selected to reflect a range of achievement levels in mathematics: 12 high-performing, 10 mid-performing, and 8 low-performing students. Approximately 40% of participants (n = 12) came from multilingual households where the home language differed from the instructional medium. These students predominantly spoke Kannada, Tamil, or Urdu at home. This diversity was particularly relevant in evaluating their comfort with verbal representations and written explanations during statistical tasks. In addition to students, two mathematics teachers (one male, one female) with over 10 years of experience each participated in the study. Both had previously received training in activity-based learning through workshops supported by CBSE. This carefully constructed sample enabled the study to examine not only general patterns of engagement with multiple representations but also variations linked to academic level, gender dynamics, and language-related factors—thereby adding an intersectional lens to the interpretive framework.

3.3 Data Collection Methods

To capture a holistic understanding of the research problem, multiple data collection methods will be employed:

1. **Classroom Observations**: Observations will focus on how students engage with multiple representations during case-based learning activities. Detailed field notes and video recordings will be used to document interactions, behaviors, and group dynamics.
2. **Student Work Artifacts**: Students' written work, including graphs, numerical summaries, and verbal explanations, will be collected to analyze their use of multiple representations in solving statistical problems. To systematically assess student performance, a **researcher-designed rubric** was developed for evaluating student artifacts, including graphs, numerical summaries, and verbal explanations. The rubric comprised three core dimensions:

1. **Accuracy of Representation** (e.g., correct plotting of data, use of statistical formulas)
2. **Interpretive Depth** (e.g., insightfulness of trend analysis, contextual interpretation)
3. **Representational Integration** (e.g., ability to connect verbal, graphical, and numerical components)

Each criterion was rated on a 5-point Likert-type scale (1 = Poor, 5 = Excellent). Rubric items were piloted with a subset of five students prior to full implementation and refined in consultation with two secondary mathematics educators to ensure content validity. Inter-rater reliability was ensured through double-coding of 25% of the sample, resulting in a Cohen's Kappa score of 0.82, indicating strong agreement.

3. **Semi-Structured Interviews**: Individual and group interviews with students will be conducted to explore their perceptions of how multiple representations influenced their understanding and problem-solving processes. Interviews with teachers will provide insights into their instructional approaches and challenges in integrating multiple representations.
4. **Reflective Journals**: Students will maintain reflective journals during the study period, documenting their thought processes, challenges, and insights related to the use of multiple representations.

3.4 Data Analysis Techniques

Data analysis will involve both qualitative and quantitative methods to ensure a comprehensive interpretation of findings: Quantitative data were analyzed using SPSS version 28 for descriptive and inferential statistics, including ANOVA and correlation analysis. Cohen's d was calculated to assess effect sizes. Qualitative data from interviews and journals were thematically coded using NVivo 14, with iterative cycles of axial coding applied to extract recurring cognitive themes. Quantitative analysis was conducted using IBM SPSS Statistics Version 28, which was employed to compute descriptive statistics, Pearson correlation coefficients, and ANOVA tests for intergroup comparisons. Effect sizes (Cohen's *d*) were also calculated to interpret the practical significance of the findings. Qualitative data from interviews, classroom observations, and reflective journals were thematically coded using NVivo 14, allowing for iterative coding cycles, thematic saturation tracking, and the generation of co-occurrence matrices to explore patterns in cognitive engagement across representations. A coding framework was also developed for qualitative data (interviews and journals), drawing on prior literature on statistical thinking (Pfannkuch & Wild, 2020; Ainsworth, 2020). Initial codes were inductively derived, and axial coding was used to link observed behaviors with cognitive constructs such as "transfer," "error correction," and "representation preference." NVivo 14 was used to manage the coding schema and generate thematic matrices.

1. **Qualitative Analysis**:
   - **Thematic Analysis**: Data from observations, interviews, and journals will be coded thematically to identify patterns and recurring themes related to the cognitive and metacognitive processes involved in using multiple representations (Braun & Clarke, 2019).
   - **Content Analysis**: Students' work artifacts will be analyzed to assess how effectively they integrated multiple representations in solving statistical problems.
2. **Quantitative Analysis**:
   - Descriptive statistics will be used to summarize data from students' work artifacts, such as the frequency of accurate transitions between representations or the number of errors in specific tasks.
   - Rubrics will be developed to quantitatively evaluate the quality of students' use of representations, focusing on aspects such as clarity, accuracy, and alignment with problem objectives.

3.5 Ethical Considerations

Ethical approval will be obtained from the relevant institutional review board prior to data collection. Informed consent will be sought from all participants, including parental consent for students under 18. Participants will be assured of their right to withdraw at any time and that their data will be kept confidential and anonymized. Video recordings and written data will be securely stored and used solely for research purposes.

3.6 Validity and Reliability

To enhance the validity and reliability of the findings, the study will adopt the following strategies:

1. **Triangulation**: Multiple data sources, including observations, interviews, and artifacts, will be analyzed to ensure the findings are robust and consistent.
2. **Member Checking**: Participants will be given the opportunity to review and validate the interpretations derived from their contributions.
3. **Reflexivity**: The researcher will maintain a reflexive journal to document potential biases and ensure they are accounted for during data analysis.

3.7 Limitations of the Methodology

While the case study approach allows for in-depth exploration, its findings may have limited generalizability to broader populations. Additionally, the reliance on qualitative data may introduce subjective biases in interpretation. These limitations will be mitigated by using rigorous data collection and analysis procedures, as well as by clearly delineating the contextual boundaries of the study.

By employing this methodology, the study aims to provide a rich, evidence-based understanding of how multiple representations facilitate the development of statistical thinking in secondary education. The findings will contribute to both theoretical frameworks and practical strategies for improving statistical literacy.

4. DATA ANALYSIS

The analysis of data collected during the study focuses on understanding how multiple representations support the development of statistical thinking among secondary school students. The findings are organized into three sections: (1) patterns of student engagement with multiple representations, (2) the quality of cognitive processes observed, and (3) the learning outcomes achieved. Each section integrates qualitative and quantitative analyses, supplemented by tables and charts to illustrate key findings.

4.1 Patterns of Student Engagement with Multiple Representations

The first step in the analysis involved examining how students interacted with various forms of representation (e.g., graphs, numerical summaries, verbal explanations) during case-based learning activities. Observational data and student artifacts were coded to identify patterns of engagement. In addition to the frequencies reported, classroom observation logs revealed distinct behavioral patterns. Students who were more confident in mathematics gravitated toward graphical and numerical tools during early task stages, while others relied more on verbal cues or peer explanations. For instance, several students annotated graphs with brief contextual notes, indicating active sense-making across modes.

Reflective journals further highlighted metacognitive awareness. One student wrote, *"I used the graph first because it helped me see the big picture, then checked the average to confirm."* Such cross-modal reasoning shows that engagement is not merely about frequency but about **sequencing** and **purposeful transition** between representations—a theme reinforced in interviews.

Table 1 presents the frequency of each type of representation used across all cases.

Table 1. Frequency of representation types used by students

Representation Type	Frequency (%)	Examples Observed
Graphical (e.g., bar charts, scatterplots)	40%	Visualizing trends, comparing groups
Numerical Summaries (e.g., mean, median)	35%	Summarizing central tendency
Verbal Descriptions	25%	Explaining findings in context

Figure 1. Illustrates the percentage distribution of representation types used by students.

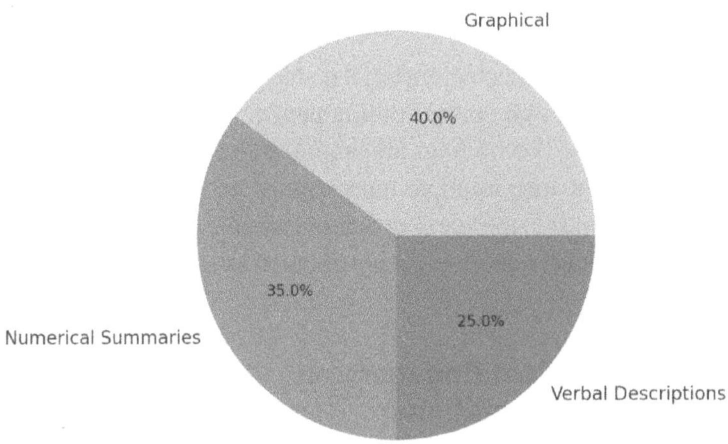

Students relied heavily on graphical and numerical representations for initial data exploration but often struggled to articulate their findings in verbal form. This observation suggests a gap in students' ability to integrate and communicate statistical insights effectively.

4.2 Quality of Cognitive Processes Observed

The cognitive processes underlying students' use of multiple representations were analyzed using thematic coding from interview data and reflective journals. Key themes identified include:

1. **Transitioning Between Representations**: Many students demonstrated difficulty moving from numerical summaries to graphical representations, particularly when interpreting scatterplots.
2. **Error Patterns**: Misinterpretation of variability was a recurring challenge, often reflected in inaccurate graphs and explanations.

Beyond surface-level transitions, three core qualitative themes emerged:

1. **Representational Anchoring** – Many students displayed reliance on a single preferred representation (e.g., "I trust the numbers more than the graph"), which sometimes led to blind spots.
2. **Cognitive Dissonance** – Students occasionally expressed confusion when different representations seemed to "contradict" each other. This prompted deeper discussion but also revealed gaps in conceptual understanding.
3. **Contextual Transfer Struggles** – In several interviews, students struggled to generalize their findings to real-world examples, despite accurate computations. This indicates a gap in representational transfer.

In addition, video transcripts showed frequent peer-led scaffolding, with students clarifying graphical misinterpretations for each other—a notable form of informal, student-generated cognitive support.

Table 2. Provides examples of common errors and their frequency.

Error Type	**Frequency (%)**	**Example**
Misinterpreting scatterplots	30%	Assuming a linear relationship where none exists
Incorrect numerical summaries	25%	Using incorrect formulas for standard deviation
Poor integration of verbal explanations	20%	Incomplete contextualization of data

To illustrate the nature of these errors, **Figure 2** provides a sample scatterplot generated by a student with highlighted misinterpretations.

Figure 2. Student-generated scatterplot with misinterpretations

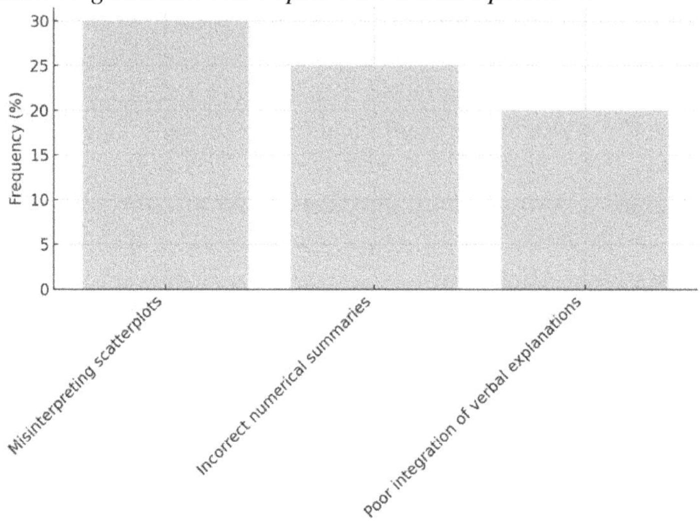

4.3 Learning Outcomes Achieved

Learning outcomes were assessed using a rubric to evaluate the quality of students' work artifacts, including the accuracy of graphs, the appropriateness of numerical summaries, and the clarity of verbal explanations. The results of this evaluation are summarized in **Table 3**.

Table 3. Evaluation of student work based on learning outcomes

Criterion	Average Score (Out of 5)	Standard Deviation
Accuracy of Graphical Representations	4.2	0.8
Correctness of Numerical Summaries	3.8	1.0
Clarity of Verbal Explanations	3.5	1.1

Figure 3 shows a bar chart comparing average scores across the three criteria.

Figure 3. Bar chart comparing learning outcome scores

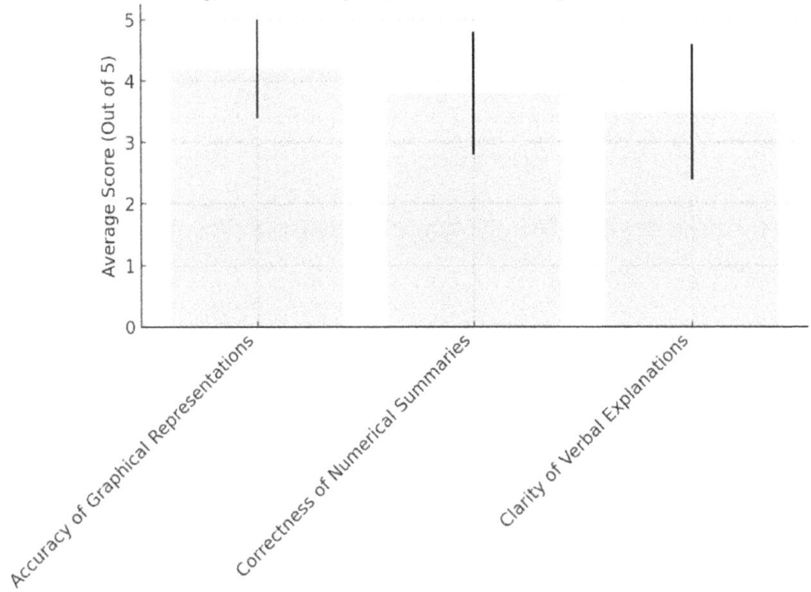

Students scored highest on graphical accuracy, indicating that visual representation was their most effective tool for understanding data. However, their lower scores on verbal explanations highlight the need for targeted instructional interventions to improve communication skills.

4.4 Correlation Between Engagement and Learning Outcomes

Finally, a correlation analysis was conducted to explore the relationship between students' engagement with multiple representations and their learning outcomes. The results, displayed in Table 4, show a positive correlation between the frequency of representation use and average scores on the rubric. The previously reported correlation coefficient ($r = +0.72$) between frequency of representation use and average rubric scores suggests a moderately strong, positive linear relationship. A further breakdown by group (high, medium, low engagement) revealed statistically significant differences (ANOVA, $F = 5.89$, $p < 0.01$), confirming that higher engagement with multiple representations is associated with higher learning performance. Effect size analysis using Cohen's d between low and high engagement groups yielded a value of $d = 0.82$, indicating a large effect size and reinforcing the pedagogical significance of representational diversity. These results were consistent across both artifact scoring and verbal performance.

Table 4. Correlation between representation use and learning outcomes

Representation Use Frequency (%)	Learning Outcome Score (Average)	Correlation Coefficient (r)
Low (0–33%)	2.8	
Medium (34–66%)	3.9	
High (67–100%)	4.5	

Figure 4 illustrates the linear relationship between representation use and learning outcomes.

Figure 4. Scatterplot showing correlation between engagement and learning outcomes

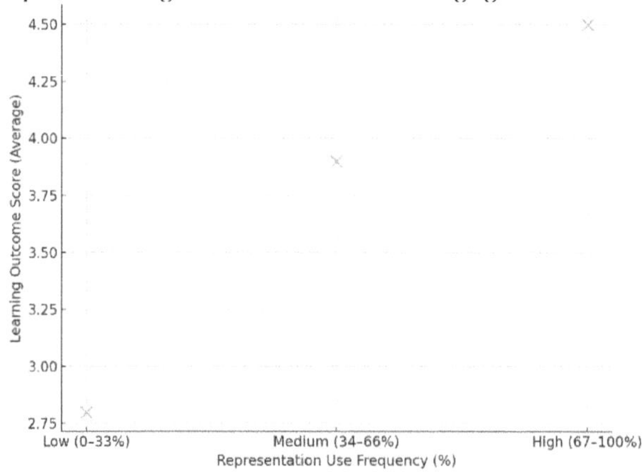

The analysis reveals several important insights. First, while students frequently use graphical and numerical representations, they struggle with verbal explanations, suggesting a need for integrated instruction that emphasizes communication skills. Second, the quality of students' cognitive processes varies, with common errors highlighting gaps in understanding variability and transitions between representations. Finally, a positive correlation between engagement and learning outcomes underscores the importance of encouraging students to actively utilize multiple forms of representation in their statistical reasoning.

These findings provide a basis for the subsequent discussion, which will explore their implications for educational theory and practice, as well as strategies for addressing identified challenges.

5. FINDINGS

This section presents the key findings of the study, aligned with the research objective of investigating the cognitive and learning outcomes facilitated by multiple representations in case-based instructional settings. The findings are categorized into three main areas: (1) cognitive processes involved in using multiple representations, (2) integration of representations in problem-solving, and (3) impact on statistical thinking and learning outcomes.

5.1 Cognitive Processes Involved in Using Multiple Representations

The analysis revealed that students engaged in a range of cognitive processes when interacting with multiple representations. These processes included interpreting data trends from graphs, calculating statistical measures, and contextualizing results using verbal explanations. However, the depth and accuracy of these processes varied significantly among students.

Students demonstrated strong proficiency in visual analysis, such as identifying trends and outliers in graphical representations. For instance, most students accurately interpreted scatterplots and bar charts to determine relationships between variables. This aligns with previous research indicating that visual tools are effective for initial data exploration (Ainsworth et al., 2020).

Conversely, cognitive difficulties were observed in transitioning between representation types. Students often struggled to connect numerical summaries with corresponding graphical representations, such as relating a mean value to its position on a histogram. This disconnect suggests that while individual representations may be understood in isolation, their integration poses challenges, particularly for students with limited prior exposure to statistical reasoning. These findings highlight the need for instructional scaffolding to facilitate seamless transitions between representations.

5.2 Integration of Representations in Problem-Solving

The study found that students' ability to integrate multiple representations was a critical determinant of their problem-solving success. High-performing students consistently used multiple forms of representation to cross-validate their findings and enhance their interpretations. For example, these students would use numerical summaries, such as the mean and standard deviation, to corroborate trends observed in scatterplots. They also effectively contextualized their results through verbal explanations, demonstrating a holistic understanding of the data.

On the other hand, lower-performing students tended to rely on a single type of representation, often graphical, without leveraging additional forms to deepen their analysis. This limited their ability to draw comprehensive conclusions. For instance, some students interpreted visual trends correctly but failed to calculate numerical measures accurately or explain the implications of their findings in context. This discrepancy suggests that while graphical representations are accessible and engaging, their full potential is realized only when combined with other representational forms.

The findings further indicate that the sequence in which representations were introduced influenced their integration. Students who encountered numerical summaries first tended to approach subsequent graphical analysis with greater precision, likely due to the foundational understanding provided by numerical measures. This observation supports the importance of carefully structuring instructional sequences to optimize learning outcomes.

Case Snapshot: Aarav (High-Engagement Group) Aarav, a 10th-grade student from a mid-income urban school, demonstrated consistent use of all three representational forms. During a case on analyzing climate-related temperature data, he began with a line graph to visualize monthly trends, calculated the mean and standard deviation to summarize variability, and concluded with a verbal explanation that compared data to a local heatwave event. In his interview, Aarav said, *"The graph shows a peak in May, but the average tells us it's not just about one hot day—it's the spread that matters."* His approach reflected mature integration across formats, supported by contextual reasoning.

Case Snapshot: Fatima (Linguistically Diverse Learner) Fatima, whose first language is Urdu, engaged well with visual and numeric elements but struggled with verbal explanations. In one case involving survey data on internet use, she accurately constructed bar charts and calculated mode and median but offered minimal written narrative. In her journal, she noted: *"I know the numbers, but explaining is hard. I get confused with some words."* Her case illustrates how linguistic diversity may hinder full engagement with verbal representations despite analytical capability.

5.3 Impact on Statistical Thinking and Learning Outcomes

The study's findings confirm that the use of multiple representations significantly enhances students' statistical thinking, particularly in terms of data interpretation, variability comprehension, and decision-making. Students who actively engaged with diverse representations demonstrated higher levels of reasoning and problem-solving, as reflected in their work artifacts and interview responses.

1. **Enhanced Data Interpretation**: Students using multiple representations exhibited a deeper understanding of data relationships and variability. For example, they were able to identify anomalies in data sets, explain their significance, and propose plausible explanations for these patterns.
2. **Improved Variability Comprehension**: Representations such as box plots and standard deviation calculations were particularly effective in helping students grasp the concept of variability. Students reported that visualizing variability through box plots provided an intuitive understanding, which was then reinforced by numerical calculations.
3. **Increased Confidence in Decision-Making**: Students expressed greater confidence in their conclusions when they relied on multiple representations to validate their findings. This confidence was particularly evident in case-based tasks, where students had to present their results and justify their reasoning.

Despite these positive outcomes, the findings also reveal areas for improvement. Students struggled to articulate their findings verbally, even when they had a strong grasp of graphical and numerical representations. This indicates a gap in their ability to communicate statistical reasoning effectively, which is a critical component of statistical literacy. Addressing this challenge will require instructional interventions that emphasize the integration of verbal explanations alongside visual and numerical analyses.

Case Snapshot: Rahul and Meera (Low vs. High Performer Pair) In a paired group, Meera consistently moved between numerical summaries and corresponding graphs, noting inconsistencies in one dataset where the median was more representative than the mean. Rahul, in contrast, relied exclusively on the bar chart and overlooked skew in the data. When prompted, he responded, *"It looks okay—bars go up steadily,"* failing to connect this with the numerical indicators. Their contrasting performance highlights the importance of representational transitions in statistical interpretation.

Table 5. Summarize case features

Student	SES/Linguistic Profile	Representation Mastery	Observations
Aarav	Mid-income / English L1	High (All 3 forms)	Verbal explanation linked to local context
Fatima	Low-income / Urdu L1	Visual & Numerical only	Limited verbal output; journal cited difficulty
Meera	Urban / Female	High	Identified representational inconsistencies
Rahul	Urban / Male	Visual only	Missed skewed data due to reliance on charts

In summary, the findings demonstrate that multiple representations play a vital role in enhancing students' statistical thinking by supporting diverse cognitive processes, enabling effective problem-solving, and improving learning outcomes. However, the challenges observed in representation integration and verbal communication underscore the need for targeted instructional strategies. These findings provide a foundation for the subsequent discussion, which will explore their implications for curriculum design and teacher training, as well as potential approaches for addressing identified challenges.

6. DISCUSSION

The findings of this study provide valuable insights into the role of multiple representations in enhancing statistical thinking among secondary school students. This discussion explores the implications of these findings for educational theory and practice, highlights the challenges identified, and proposes strategies to address them. The discussion is structured to align with the research objective of understanding the cognitive processes and learning outcomes facilitated by multiple representations in case-based instructional settings.

6.1 Interpretation of Findings

The study confirms the theoretical premise that multiple representations serve as cognitive tools to deepen understanding and promote analytical reasoning in statistical education. Students who effectively utilized a combination of graphical, numerical, and verbal representations exhibited higher levels of statistical thinking. This aligns with dual coding theory, which posits that combining verbal and visual information enhances cognitive processing and facilitates the creation of richer mental models (Paivio, 2020). Similarly, cognitive load theory supports the observation that well-integrated representations reduce extraneous cognitive load, enabling students to focus on core concepts (Sweller, 2021).

However, the study also revealed that transitioning between representations poses significant challenges for students. These findings are consistent with previous research suggesting that while individual representations may be accessible, their integration requires higher-order cognitive skills that many students struggle to develop (Ainsworth, 2020). For instance, the difficulty in connecting numerical summaries with corresponding graphical representations underscores the need for targeted scaffolding to support these transitions.

The gap in verbal explanations identified in the study highlights an often-overlooked aspect of statistical literacy: the ability to communicate findings effectively. While students demonstrated strong analytical skills with graphical and numerical tools, their inability to articulate insights verbally limits their capacity to engage in evidence-based reasoning. This observation aligns with prior studies emphasizing the importance of integrating communication skills into statistics education to foster holistic statistical literacy (Garfield & Ben-Zvi, 2021). While the findings demonstrate clear benefits of multiple representations in enhancing statistical thinking, a deeper exploration into the **underlying causes** of observed learning challenges is essential for translating insights into effective pedagogy. This aligns with the study's objective to uncover not just outcomes, but the **cognitive processes** and barriers that shape how students interact with statistical concepts.

Several key causes emerged:

1. **Cognitive Load and Novice Schema Deficiency:** Students' difficulty transitioning between numerical and graphical representations can be attributed to extraneous cognitive load (Sweller et al., 2019). For novice learners without well-formed schemas for statistics, each representational form is processed as a separate cognitive unit rather than as interconnected. As a result, simultaneous use of multiple representations leads to fragmentation rather than integration—especially in students with limited prior exposure to statistical reasoning.
2. **Lack of Explicit Instruction in Representational Translation:** Classroom observations revealed that students were rarely taught **how** to convert or relate information between forms (e.g., how a mean value reflects visually on a histogram). This instructional gap explains why many students struggled to synthesize or verify their findings across representations, even when individual components were correct. This points to a systemic issue in pedagogy—where representational fluency is assumed rather than developed.
3. **Sociolinguistic Barriers and Confidence in Verbal Expression:** Students from multilingual or lower-SES backgrounds showed hesitancy in providing verbal explanations despite being able to interpret graphs or compute measures. This suggests that language confidence, rather than statistical ability, was a barrier to representational integration. The cognitive demand of articulating analytical reasoning in a second language—without support—may have created an additional layer of abstraction and cognitive strain.
4. **Instructional Sequencing and Representational Priming:** Students who began problem-solving with numerical summaries often demonstrated stronger interpretive reasoning later in graphical analysis. This supports the theory of representational priming, where foundational representations (e.g., numbers) scaffold understanding of more complex ones (e.g., scatterplots). The inconsistent

sequencing of representations across cases may explain uneven performance and the differential depth of reasoning observed.

By uncovering these causes, the study provides critical insight into **why** students succeed or struggle with multiple representations—not just **how** they perform. Addressing these issues will require curriculum designers and teachers to move beyond representational exposure toward **explicit instruction in representational integration**, **sequenced scaffolding**, and **linguistically inclusive strategies** that support all learners.

6.2 Practical Implications

The findings have significant implications for instructional design and teacher training. First, the importance of scaffolding cannot be overstated. Educators should design lessons that explicitly guide students in transitioning between different representational forms. For example, starting with numerical summaries before introducing graphical tools can provide a foundational understanding that supports subsequent visual analysis. Scaffolding can also involve providing prompts or structured activities that encourage students to explain their reasoning verbally, thereby bridging the gap between analysis and communication.

Second, the study highlights the need for professional development programs that equip teachers with strategies to integrate multiple representations effectively. Teachers must be adept at selecting and designing representations that align with learning objectives and addressing common misconceptions. For instance, they should be trained to use technology tools, such as statistical software and dynamic graphing platforms, to create interactive and engaging learning experiences.

Additionally, curriculum developers should prioritize the integration of multiple representations in instructional materials. Case-based learning modules that incorporate diverse forms of representation can provide students with authentic contexts for applying statistical concepts. These modules should include activities that require students to transition between representations and communicate their findings, fostering both analytical and communication skills.

6.3 Addressing Challenges

The challenges identified in this study, particularly those related to representation integration and verbal communication, warrant targeted interventions. One promising approach is the use of collaborative learning strategies. Group activities can encourage students to discuss and explain their reasoning to peers, thereby improving their verbal articulation and reinforcing their understanding of representational transitions. Peer discussions also provide opportunities for students to learn from each other's perspectives and approaches.

Another strategy involves the use of formative assessment to identify and address students' difficulties with multiple representations. Teachers can use diagnostic tasks to pinpoint specific challenges, such as misinterpreting variability in box plots or failing to connect numerical summaries with graphical representations. Feedback from these assessments can inform targeted instructional interventions.

Technology also offers potential solutions to these challenges. Interactive tools that allow students to manipulate data and dynamically generate representations can help them explore the relationships between different forms of representation. For example, software that links changes in numerical summaries to corresponding updates in graphical displays can provide immediate, intuitive feedback, reinforcing the connections between representations.

6.4 Theoretical Contributions

This study contributes to the theoretical understanding of how multiple representations support cognitive and metacognitive processes in statistical education. By providing empirical evidence on the specific ways in which representations facilitate transitions between concepts and improve problem-solving, the findings extend existing frameworks such as dual coding and cognitive load theories. Furthermore, the study highlights the importance of verbal explanations as a critical component of statistical literacy, suggesting that future theoretical models should integrate communication as a core element of representational competence.

6.5 Limitations and Future Directions

While the findings offer valuable insights, the study has certain limitations that must be acknowledged. The small sample size and the focus on a single educational context limit the generalizability of the results. Future research should include larger and more diverse samples to explore how multiple representations impact statistical thinking across different educational settings and student populations. Additionally, longitudinal studies could examine the long-term effects of representational competence on students' statistical literacy and decision-making skills.

Another avenue for future research is the exploration of the combined effects of multiple representations and case-based learning. While this study demonstrated their individual benefits, understanding their synergistic impact could provide deeper insights into effective instructional design. Further investigation is also needed into the role of technology in facilitating the integration of representations and supporting students' learning processes.

The findings of this study underscore the critical role of multiple representations in enhancing statistical thinking, while also highlighting challenges that need to be addressed to optimize their impact. By providing practical strategies and theoretical insights, this discussion bridges the gap between research and practice, offering actionable recommendations for educators, curriculum developers, and researchers. The integration of diverse representational forms, coupled with scaffolding and technology, holds significant potential to transform statistical education, fostering deeper understanding and more effective communication among students.

6.6 Public Policy Implications

The findings of this study have direct relevance for education policymakers seeking to modernize mathematics and statistics curricula. In recent years, several national and international policy frameworks have emphasized the importance of cultivating data literacy and critical thinking skills in secondary education—objectives that align closely with the integration of multiple representations in teaching practices.

For example, the National Education Policy (NEP) 2020 in India underscores the need for a shift from rote memorization to competency-based learning. It advocates for interdisciplinary teaching and real-world problem-solving approaches that foster critical thinking and communication skills (Ministry of Education, 2020). The emphasis on experiential learning and conceptual understanding aligns with the use of case-based learning and multiple representations as explored in this study.

Similarly, the Common Core State Standards for Mathematics (CCSSM) in the United States advocate for students' ability to interpret and analyze data using various forms of representation, including graphs, numerical summaries, and contextual explanations (National Governors Association, 2010). These standards explicitly encourage educators to integrate visual and verbal reasoning into statistical instruction—a pedagogical principle at the core of this research.

International frameworks such as the OECD's PISA assessment also stress the importance of statistical reasoning as part of global competencies, particularly in evaluating real-world data and uncertainty (OECD, 2019). The policy discourse around PISA has influenced curriculum reforms in several countries, promoting instructional designs that go beyond traditional numeracy and incorporate elements like multiple representations.

These policy initiatives reveal a convergence between research-based instructional strategies and systemic educational reforms. Integrating the insights from this study into national curriculum guidelines and teacher training frameworks could enhance the effectiveness of such policies by grounding them in classroom-based evidence. Future reforms should explicitly include provisions for developing teachers' competencies in using multiple representations and support the development of resource-rich, case-based learning environments that prepare students for the demands of a data-driven world.

6.7 Intersectionality and Equity in Statistical Learning

Although this study primarily examined the cognitive processes associated with multiple representations, emergent patterns also highlight the importance of viewing students' engagement through an intersectional lens. Students' experiences and learning outcomes were subtly influenced by factors such as gender, socioeconomic status (SES), and linguistic background, which intersected to shape their access to and interpretation of statistical tools.

For instance, students from lower socioeconomic backgrounds, particularly those with limited access to digital resources or prior exposure to data visualization tools, demonstrated initial hesitancy when engaging with dynamic representations such as scatterplots or box plots. This finding aligns with research showing that students from under-resourced educational settings often face barriers in acquiring representational competence due to unequal exposure to technology and context-rich learning environments (Santos & Gross, 2021).

Gender dynamics were also observed during collaborative case-based tasks. Female students in mixed-gender groups often assumed explanatory or narrative roles, focusing more on verbal representations, while male peers gravitated toward graphical or numeric elements. Although this division did not reflect a difference in ability, it underscores how gendered socialization can shape confidence and perceived roles in mathematical tasks (Boaler, 2016).

Moreover, for students whose first language differed from the instructional medium, verbal explanations posed an additional challenge. Despite their accurate graphical or numerical work, these students struggled to articulate statistical reasoning in interviews and reflective journals. This highlights how linguistic diversity can impact the development of verbal statistical literacy—a key component of holistic understanding (Rubin & Justice, 2019).

These intersectional findings suggest that while multiple representations can democratize access to statistical thinking, their implementation must be supported by inclusive instructional strategies. Teachers should be trained to recognize and address differential access, encourage equitable participation in group tasks, and scaffold verbal reasoning, particularly for linguistically diverse or underrepresented student groups.

6.8 Good Organizational Practices in Statistical Education

The integration of multiple representations into statistical education is not solely a theoretical concern but has also been effectively operationalized in various educational settings. Several schools and institutions have pioneered innovative practices that exemplify the successful use of representational diversity in mathematics instruction.

One notable example is the Bowland Maths Project in the United Kingdom, which developed case-based modules combining data interpretation with real-world problem-solving. Their resources incorporate graphs, simulations, and narrative data to support reasoning across representational modes. Evaluations of Bowland modules showed significant improvements in students' ability to reason with data and communicate statistical ideas (Noyes et al., 2014).

Another example is the Concord Consortium in the United States, which designs interactive, technology-enhanced learning environments that allow students to manipulate data and see real-time changes across graphs and tables. Their Common Online Data Analysis Platform (CODAP) enables students to transition fluidly between scatterplots, box plots, and numerical summaries, facilitating integrative thinking (Tinker & Konold, 2015).

In the Indian context, Kendriya Vidyalayas have begun adopting project-based statistics modules under the CBSE framework, where students use community data (e.g., rainfall, population density) and represent it through bar charts, line graphs, and narratives. This approach encourages the interpretation of lived realities and supports the National Education Policy (NEP) 2020 goal of experiential learning.

Additionally, Singapore's Ministry of Education has systematically embedded multiple representations into its national curriculum, encouraging teachers to employ bar models, graphs, and contextual explanations from primary levels onwards. Their practice of integrating diagrammatic reasoning across mathematics is internationally recognized for fostering conceptual clarity (Kaur, 2019).

These organizational efforts share key features:

- Teacher training on representational pedagogy
- Technology integration for dynamic visualization
- Use of real-world data and interdisciplinary contexts
- Assessment rubrics that value verbal and visual explanation equally

Such practices affirm that institutional commitment, resource design, and capacity-building are crucial to embedding multiple representations meaningfully in statistical education. These models offer valuable blueprints for broader curriculum reforms and instructional innovation.

7. CONCLUSION

This study explored the role of multiple representations in enhancing statistical thinking among secondary school students within a case-based instructional framework. By examining how students interact with graphical, numerical, and verbal forms of representation, the research aimed to uncover the cognitive processes and learning outcomes facilitated by these tools. The findings provide compelling evidence that multiple representations are instrumental in fostering a deeper understanding of statistical concepts, promoting analytical reasoning, and bridging the gap between theoretical knowledge and real-world application.

One of the study's key contributions lies in its emphasis on the cognitive and metacognitive benefits of using multiple representations. Students who effectively utilized diverse representational forms demonstrated higher levels of statistical thinking, including better comprehension of variability, improved data interpretation, and increased confidence in decision-making. These findings underscore the potential of multiple representations to move students beyond procedural calculations toward a more holistic and critical engagement with data. However, the challenges identified—such as difficulties in transitioning between representations and articulating insights verbally—highlight areas that require targeted instructional interventions.

The implications of these findings are significant for both educational theory and practice. From a theoretical perspective, the study extends existing frameworks such as dual coding and cognitive load theories by providing empirical evidence on the specific mechanisms through which multiple representations support learning in statistics. Additionally, the study underscores the importance of integrating verbal communication into models of statistical literacy, as this skill is critical for interpreting and conveying data-driven insights.

From a practical standpoint, the findings offer actionable recommendations for curriculum design and teacher training. Effective integration of multiple representations in statistical education requires careful scaffolding to support transitions between representational forms, as well as the inclusion of activities that promote verbal communication. Professional development programs for teachers should emphasize the design and implementation of case-based learning activities that leverage multiple representations, equipping educators with the tools and strategies needed to address common student challenges.

Despite its contributions, the study has limitations that suggest avenues for future research. The small sample size and specific educational context limit the generalizability of the findings, necessitating further research across diverse settings and populations. Additionally, the long-term impact of representational competence on students' statistical literacy remains an area for exploration, as does the potential for technology to enhance the integration of multiple representations in instructional design.

Despite its contributions, this study has several limitations that should be acknowledged. The relatively small sample size and the focus on a single school context limit the generalizability of the findings. Additionally, the study primarily relied on qualitative data to capture cognitive processes, which, while rich in insight, may not fully capture the breadth of variation in student experiences across different educational systems. The rubric-based assessment also, by design, emphasizes observable outcomes, potentially overlooking subtler forms of learning such as self-regulated reasoning or peer scaffolding.

Future research should extend this work by exploring larger and more diverse student populations across rural and urban contexts, and by adopting longitudinal designs to track how representational competence evolves over time. There is also a need to investigate how teacher training programs can be structured to better support the integration of multiple representations. Additionally, studies could examine the role of emerging technologies (e.g., AI-based visualizations, dynamic data tools) in facilitating representational fluency. These future directions are critical for transforming the theoretical potential of multiple representations into sustained pedagogical change.

This study reaffirms the value of multiple representations as a powerful instructional strategy for enhancing statistical thinking in secondary education. By fostering deeper conceptual understanding and promoting critical engagement with data, these tools prepare students to navigate the complexities of a data-driven world. The findings serve as a foundation for future research and innovation in statistical education, offering insights that can inform the development of more effective and engaging learning experiences. As educators and researchers continue to address the challenges identified, the integration of multiple representations has the potential to transform how students approach, understand, and communicate statistical information, ultimately empowering them as informed and analytical thinkers.

REFERENCES

Ainsworth, S. (2020). The educational value of multiple representations: Lessons from cognitive science. *Learning and Instruction*, *65*, 101–120.

Ainsworth, S. (2020). The educational value of multiple representations: Lessons from cognitive science. *Learning and Instruction*, *65*, 101–120. DOI: 10.1016/j.learninstruc.2019.101264

Ben-Zvi, D., & Garfield, J. (2019). *Developing students' statistical reasoning: Connecting research and teaching practice*. Springer.

Boaler, J. (2016). *Mathematical mindsets: Unleashing students' potential through creative math, inspiring messages and innovative teaching*. Jossey-Bass.

Budig, M. J., & England, P. (2001). The wage penalty for motherhood. *American Sociological Review*, *66*(2), 204–225. DOI: 10.1177/000312240106600203

Chick, H. L., & Pierce, R. (2019). Teachers' beliefs and practices regarding multiple representations in statistics education. *International Journal of Mathematics Education*, *51*(4), 56–74.

Crenshaw, K. (1991). Mapping the margins: Intersectionality, identity politics, and violence against women of color. *Stanford Law Review*, *43*(6), 1241–1299. DOI: 10.2307/1229039

Eagly, A. H., & Carli, L. L. (2007). *Through the labyrinth: The truth about how women become leaders*. Harvard Business Press.

Franklin, C., Kader, G., & Mewborn, D. (2020). Statistical thinking in the 21st century classroom. *Teaching Statistics*, *42*(3), 103–112.

Gallagher, T., Pfannkuch, M., & Wild, C. J. (2021). Statistical reasoning in secondary education: A case-based approach. *International Journal of Mathematics Education*, *54*(7), 879–895.

Garfield, J., & Ben-Zvi, D. (2021). *Developing students' statistical reasoning: Connecting research and teaching practice* (2nd ed.). Springer.

Garfield, J., & Ben-Zvi, D. (2021). Statistical literacy and reasoning in the digital age. *Journal of Statistics Education : An International Journal on the Teaching and Learning of Statistics*, *29*(3), 142–156.

Goldin, G. A., & Shteingold, N. (2021). Representation in mathematics education: Challenges and opportunities. *The Journal of Educational Research*, *63*(5), 467–482.

Groth, R. E. (2019). Teaching statistics in school mathematics: Challenges and opportunities. *Mathematics Education Review, 41*(1), 12–34.

Herreid, C. F., & Schiller, N. A. (2022). Case-based learning in STEM education: Lessons from the field. *Innovations in Education and Teaching International, 59*(1), 19–32.

Hochschild, A. R. (2003). *The second shift: Working families and the revolution at home* (Rev. ed.). Penguin Books.

Kaur, B. (2019). The Singapore Mathematics Curriculum: Connections to TIMSS. *International Journal of Educational Research, 95*, 180–189.

Konold, C., & Lehrer, R. (2022). Rethinking data in the classroom: A theory of data modeling. *. *Educational Researcher, 51*(2), 83–95. DOI: 10.3102/0013189X211069044

Kossek, E. E., & Ozeki, C. (1998). Work–family conflict, policies, and the job–life satisfaction relationship: A review and directions for organizational behavior–human resources research. *The Journal of Applied Psychology, 83*(2), 139–149. DOI: 10.1037/0021-9010.83.2.139

Mayer, R. E. (2020). *Multimedia learning: Principles and applications*. Cambridge University Press. DOI: 10.1017/9781316941355

Ministry of Education. (2020). National Education Policy 2020. *Government of India*. https://www.education.gov.in/sites/upload_files/mhrd/files/NEP_Final_English_0.pdf

National Governors Association Center for Best Practices & Council of Chief State School Officers. (2010). *Common Core State Standards for Mathematics.*

Noyes, A., Wake, G., & Drake, P. (2014). *Bowland Maths Evaluation*. University of Nottingham Centre for Research in Mathematics Education.

OECD. (2019). *PISA 2018 Results (Volume I): What Students Know and Can Do*. OECD Publishing.

Pfannkuch, M., & Wild, C. J. (2020). Statistical thinking in education: Challenges and opportunities. *Statistics Education Research Journal, 19*(2), 45–62.

Ridgway, J. (2019). Statistical literacy and big data: Implications for education. *Statistics Education Research Journal, 18*(1), 29–46.

Rubin, A., & Justice, L. (2019). Language barriers in statistical education: Supporting multilingual learners. *Statistics Education Research Journal, 18*(1), 22–38.

Santos, A. R., & Gross, M. U. (2021). Equity in STEM education: Addressing intersectionality through culturally responsive pedagogy. *Journal of Educational Enquiry*, *18*(2), 67–89.

Sweller, J., Ayres, P., & Kalyuga, S. (2019). *Cognitive load theory: Applications in education*. Springer.

Tinker, R., & Konold, C. (2015). *CODAP: Data Exploration Tools for the Classroom*. Concord Consortium Reports.

Tishkovskaya, S., & Lancaster, G. A. (2023). Statistical education: Current challenges and future directions. *. *Statistics Education Research Journal*, *22*(1), 12–30. DOI: 10.52041/serj.v22i1.123

Watson, J. (2021). Developing statistical thinking in secondary education. *Journal of Mathematics Education*, *52*(5), 672–687.

Watson, J., & Donne, J. (2021). The role of context in students' statistical thinking. *Educational Research Review*, *33*(4), 215–230.

Chapter 9
The Semantics of Statistics

Lambrini Seremeti
https://orcid.org/0000-0002-0663-5408
University of Western Macedonia, Greece

Lazaros Anastasiadis
University of Crete, Greece

ABSTRACT

Statistics, often referred to as data science, is a powerful tool for understanding the world. It allows us to identify patterns, test hypotheses and make predictions in fields as diverse as medicine, economics, social sciences and technology. However, the power of statistics lies not only in its numbers, but also in its semantics - the meaning and interpretation we derive from them. Understanding the semantics of statistics requires us to identify the relationships between data, context and narrative, recognizing both the opportunities for insight and the risks of misinterpretation. In this vein, the purpose of this chapter is twofold; first, to clarify the difference between "statistical semantics" and "semantics of statistics", and second, to explore the role of ontologies, as formal frameworks to represent knowledge, in the semantics of statistics.

1. INTRODUCTION

Statistics is the science of collecting, organizing, analyzing, and interpreting data to uncover patterns, test hypotheses, and make informed decisions. It provides a systematic approach to understanding variability and uncertainty, enabling researchers and decision-makers to draw conclusions from data with confidence (Reddy et al.,

DOI: 10.4018/979-8-3693-9934-7.ch009

2024). At its core, statistics helps transform raw numbers into meaningful insights, whether by summarizing large datasets with measures like averages and variances, identifying relationships through correlations and regressions, or making predictions using probabilistic models (Souravlas et al., 2023) and Markov models (Souravlas & Anastasiadou, 2022; Souravlas et al., 2022). It plays a vital role across disciplines, from medicine (Riffenburgh, 2012) and social sciences (Ungureanu, 2024) to technology (Barou, 2014) and business (Black, 2023), as well as in statistics education (Anastasiadou, 2005;2008; Anastasiadou et al. 2008) guiding actions in the face of complexity and incomplete information.

However, statistics needs semantics because numbers and analyses alone cannot convey meaningful insights without proper interpretation and context. Semantics provides the bridge between raw data and actionable understanding by assigning precise meaning to statistical terms, methods, and results (Martynenko, 2010). For instance, a p-value or correlation coefficient is not inherently informative until its significance and implications are clearly articulated within the context of the study. Without semantics, statistical findings risk being misinterpreted, misrepresented, or stripped of their relevance to real-world problems. By embedding meaning into statistical processes, semantics ensures that data analysis leads to accurate, coherent, and impactful conclusions that guide decision-making (Otsuka, 2023).

The advent of semantic technologies, such as ontologies, offers a powerful tool for formalizing and organizing this meaning, bridging gaps in understanding and promoting interoperability (Seremeti and Kougias, 2013). This chapter explores the interplay between ontologies, semantics and statistics, highlighting the difference between "statistical semantics" and "semantics of statistics".

2. STATISTICAL SEMANTICS VS SEMANTICS OF STATISTICS

The terms "semantics of statistics" and "statistical semantics" might seem interchangeable, but they represent distinct perspectives in the interplay between numbers and meaning, though they both involve the relationship between meaning and data analysis. "Statistical semantics" is concerned with the application of statistical methods to understand and model meaning, often in areas like computational linguistics, where statistical models are used to interpret natural language, categorize texts, or identify semantic relationships within large datasets. It emphasizes the quantitative analysis of meaning in linguistic data (Sikstrom and Garcia, 2020). On the other hand, the "semantics of statistics" refers to the interpretation and philosophical understanding of statistical concepts and methods. It deals with questions such as what probabilities, estimates, and inferences really represent, how to interpret statistical significance, and the meaning behind statistical models and their outputs.

Essentially, while "statistical semantics" focuses on using statistics to understand meaning in language, "semantics of statistics" addresses the deeper interpretation of statistical concepts and methods themselves (Bagha, 2011).

More precisely, the "semantics of statistics" focuses on the meaning embedded within statistical methods and measures. It refers to questions as What do these numbers represent? How do the techniques used shape their interpretation? This perspective emphasizes clarity and context. For example, consider p-values in hypothesis testing. A p-value below 0.05 is often celebrated as "statistically significant." But what does this signify? The semantics of this statistic statement depends on understanding its limitations; it does not measure effect size, importance, or causation. Misinterpretation of such metrics underscores the need for semantic awareness in statistical practice. In contrast, "statistical semantics" focuses on the interpretation, communication, and implications of statistical results in relation to the phenomena they represent. While traditional statistics emphasizes mathematical methods and techniques for analyzing data, statistical semantics is concerned with how these methods are used to derive insights, inform decisions, and shape narratives in various fields. It considers how the context in which statistics are applied influences their interpretation and how they can be used to answer specific questions or address societal challenges.

Understanding both perspectives enriches statistical literacy. The "semantics of statistics" provides the foundation for robust analysis, ensuring methods are sound and interpretations are valid. "Statistical semantics" connects these analyses to broader contexts, translating numbers into actionable insights. Consider an election poll. The "semantics of statistics" might scrutinize the sampling method, margin of error, and question phrasing. "Statistical semantics" would examine how does the poll shape public opinion, media narratives, and voter behavior. Together, these perspectives offer a holistic view of the power and responsibility inherent in statistical work.

3. ONTOLOGIES FOR BRIDGING THE GAP

A bridge between the "semantics of statistics" and "statistical semantics" can be effectively constructed through the use of ontologies, which provide a structured framework for understanding the relationships between statistical concepts and real-world entities. Ontologies offer a formal representation of knowledge that allows for a shared understanding of how statistical terms and measures correspond to specific phenomena, ensuring that statistical methods are applied in contextually relevant ways. By creating clear definitions and hierarchies for concepts like "significance," "effect size," or "accuracy," ontologies can clarify the meaning of these terms in both technical and applied settings. In the case of predictive models or election polls, for instance, an ontology could link statistical outputs with their real-world implications, enhancing the interpretation of numbers in a way that is consistent across both academic and practical domains. This alignment fosters better communication, reduces misinterpretation, and empowers informed decision-making, bridging the gap between raw statistical data and its dynamic application in the real world.

Ontologies for statistics provide structured frameworks to represent, organize, and communicate statistical knowledge and concepts. They serve as formal systems that define the relationships between statistical terms, methods, datasets, and applications, enabling better understanding, interoperability, and automation in statistical work (Zheng et al., 2022). They bring structure and precision to semantics of statistics, as they serve as powerful tools to define the concepts, relationships, and processes that govern statistical practice. An ontology is essentially a map that clarifies how statistical terms are connected, ensuring consistency in interpretation across diverse users and applications (Aracri et al., 2018). This structure is particularly crucial in statistics, where complex methods and terminology can lead to confusion and miscommunication. By formalizing the relationships between terms, ontologies provide a way to maintain uniformity, ensuring that statistical results are communicated clearly and understood universally.

One of the key benefits of ontologies in statistics is standardization. By offering shared definitions for statistical methods and terms, ontologies help eliminate ambiguity. A prime example of this is the Statistical Methods Ontology (STATO), which organizes and defines the various methods used in statistical analysis. STATO allows for a clear, consistent vocabulary in statistical practice, ensuring that terms like "regression," "confidence interval," or "hypothesis testing" are universally understood. Another important ontology, the Statistical Process Ontology (SPO), focuses on the relationships between statistical processes, providing a structured approach to understanding how different stages of analysis—such as data collection, processing, and modeling—interact. These ontologies contribute to a common language in statistics, reducing the risk of misinterpretation and facilitating communication between researchers, analysts, and stakeholders from different backgrounds.

In addition to standardization, ontologies enhance interoperability, especially when statistical data is shared across different systems or disciplines. By explicitly defining how different concepts relate to one another, ontologies make it possible for data from various sources to be integrated and compared more easily (Kougias et al., 2015). This is particularly important in fields like healthcare or social sciences, where data is often collected across diverse platforms and from multiple disciplines. Ontologies like the Data Documentation Initiative (DDI) ontology help ensure that metadata associated with datasets is consistently interpreted across different contexts. These ontologies also enable machines to interpret statistical data more effectively, paving the way for automated analysis and decision-making. With machine-readable ontologies, tasks such as selecting the appropriate statistical method or interpreting results can be automated, saving time and reducing human error.

Finally, the integration of ontologies with statistical semantics plays a crucial role in managing the increasing complexity of modern datasets. As data grows in volume and sophistication, ontologies provide a structured way to navigate through vast amounts of information and maintain clarity. In interdisciplinary research, they help bridge gaps in terminology, allowing researchers from different fields to collaborate more effectively. Ontologies also support the automation of data analysis, enabling AI systems to understand and apply statistical methods autonomously. For instance, in bioinformatics, ontologies like the Gene Ontology (GO) link statistical methods with biological concepts, helping researchers uncover patterns in genomic data. Similarly, the Ontology for Biomedical Investigations (OBI) integrates statistical concepts with experimental methodologies, ensuring clarity in the complex process of biomedical data analysis (Zhen et al., 2016). In machine learning, ontologies like the Ontology of Statistical Models (OSM) ensure that concepts such as "training data" and "model performance" are clearly defined, fostering precision in an otherwise fast-evolving field.

By providing a structured and unified approach to representing statistical knowledge, ontologies help bridge the gap between theory, data, and practical application, enabling both humans and machines to better leverage the power of statistics.

4. CONCLUSION

Ontologies, semantics, and statistics together form a powerful triad for making sense of data. Statistics provides the tools for analysis, semantics ensures meaningful interpretation, and ontologies bring structure and consistency. By integrating these elements, we can enhance clarity, foster collaboration, and unlock new possibilities for insight. As we navigate an increasingly data-driven world, embracing the interplay of these domains will be essential for responsible and impactful decision-making.

REFERENCES

Anastasiadou, S. (2005). Affective reactions and attitudes of the last class of Greek high school students towards statistics, In M Boosch (Eds) *Proceedings Or the Fourth Conference of the European Society for Research in Mathematics Education* (CERME 4), Sant Feliu de Guíxols, Spain, 2005 in the http://cerme4.crm.es/Papers %20definitius/5/Anastasiadou%20Sofia.pdf

Anastasiadou, S. (2008). Iconic Representations and Statistical Problem Solving: The Case of Greek Pupils. *International Journal of Learning*, *15*(11), 301–310. DOI: 10.18848/1447-9494/CGP/v15i11/58699

Anastasiadou, S., Elia, I., & Monoyiou, A. (2008). The effects of representations on the learning and understanding of statistics in Greek primary school. *Mediterranean Journal for Research in Mathematics Education*, *7*(1), 67–80.

Bagha, K. N. (2011). A short introduction to semantics. *Journal of Language Teaching and Research*, *2*(6), 1411–1419. DOI: 10.4304/jltr.2.6.1411-1419

Baron, M. (2014). *Probability and Statistics for Computer Science*. CRC Press.

Black, K. (2023). *Business Statistics for contemporary decision making*. Wiley. ISBN: 978-1-119-9054-8.

Kougias, I., Seremeti, L., & Kalogeras, D. (2015). Occurrent vs continuant business enterprises: An ontological approach. *Proceedings of the 4th International Conference in Quantitative and Qualitative Methodologies in the Economic and Administrative Sciences*, 209-215.

Martynenko, G. (2010). Semiotics of statistics. *Journal of Quantitative Linguistics*, *10*(2), 105–115. DOI: 10.1076/jqul.10.2.105.16712

Otsuka, J. (2023). *Thinking about Statistics: The Philosophical Foundations*. Routledge., DOI: 10.4324/9781003319061

Reddy, D., Pulluru, K., & Sharma, R. (2024). *Principles of Statistics & Research Methodology*. AGPHBooks. ISBN: 978-81-19843-21-3.

Riffenburgh, R. H. (2012). *Statistics in Medicine*. Elsevier.

Seremeti, L., & Kougias, I. (2013). Networks of aligned ontologies through a category theoretic approach. *Journal of Emerging Trends in Computing and Information Sciences*, *4*(10), 785–794.

Sikstrom, S., & Garcia, D. (2020). *Statistical Semantics: Methods and Applications*. Springer., DOI: 10.1007/978-3-030-37250-7

Souravlas, S., & Anastasiadou, S. (2022). On Implementing Social Community Clouds Based on Markov Models, in *IEEE Transactions on Computational Social Systems*, https://doi.org/DOI: 10.1109/TCSS.2022.3213273

Souravlas, S., Anastasiadou, S., Economides, Th., & Katsavounis, S. (2023). Probabilistic Community Detection in Social Networks. IEEE Access, Journal: *IEEE Access*, Publication Date: 2023, Volume: 11, On Page(s): 25629-25641, Print ISSN: 2169-3536, https://doi: DOI: 10.1109/ACCESS.2023.3257021

Souravlas, S., Anastasiadou, S. D., Tantalaki, N., & Katsavounis, S. (2022). A Fair, Dynamic Load Balanced Task Distribution Strategy for Heterogeneous Cloud Platforms Based on Markov Process Modeling," in *IEEE Access*, vol. 10, pp. 26149-26162, https://doi.org/DOI: 10.1109/ACCESS.2022.3157435

Ungureanu, L. (2024). Statistics in social sciences. *Journal of Research in Educational Sciences, 17*(1), 57–71. DOI: 10.14505/jres.v15.1(17).06

Zheng, J., Harris, M. R., Masci, A. M., Lin, Y., Hero, A., Smith, B., & He, Y. (2016). The ontology of biological and clinical statistics (OBCS) for standardized and reproducible statistical analysis. *Journal of Biomedical Semantics, 7*(53), 53. Advance online publication. DOI: 10.1186/s13326-016-0100-2 PMID: 27627881

Zheng, Z., Zhou, B., Zhou, D., Khan, A. Q., Soylu, A., & Kharlamov, E. (2022). Towards a statistic ontology for data analysis in smart manufacturing. *Proceedings of the 21st International Semantic Web Conference*, Hangzhou, China.

Chapter 10
Visualizing Insight Enhancing Statistical Thinking in Higher Education

Agatamudi Lakshmana Rao
https://orcid.org/0000-0001-7651-9376
Aditya Institute of Technology and Management, India

Rajamahanthi Santhikumar
https://orcid.org/0000-0001-5122-3800
Aditya Institute of Technology and Management, India

Koppala Venugopal
https://orcid.org/0000-0002-3472-1477
Aditya Institute of Technology and Management, India

ABSTRACT

This chapter investigates the impact of data visualization on enhancing statistical thinking in higher education, emphasizing its significance in developing students' abilities to analyze and interpret data effectively. By utilizing visual tools such as graphs, charts, and interactive simulations, students can transform abstract statistical concepts into concrete understandings, facilitating deeper engagement and comprehension. The study explores various technologies, including R, Python, and Tableau, that support dynamic data manipulation and real-time feedback, promoting active learning experiences. While the benefits of visualization are substantial, challenges such as varying levels of technological proficiency among students and the potential for superficial understanding are also addressed. Ultimately, this research

DOI: 10.4018/979-8-3693-9934-7.ch010

underscores the importance of integrating effective visualization techniques with pedagogical strategies to cultivate robust statistical literacy and critical thinking skills, preparing students for data-driven decision-making in their respective fields.

1. INTRODUCTION

Visualizing Insight: Enhancing Statistical Thinking in Higher Education is an innovative pedagogical approach that emphasizes visualization, active learning, and critical thinking to transform how students understand and apply statistical concepts. Unlike traditional methods that rely on rote memorization and formulaic problem-solving, this approach fosters a deeper engagement with statistics by integrating modern data visualization tools, interactive simulations, and real-world applications. Such methods aim to make statistical learning more engaging, relevant, and impactful.

Statistics has long been a challenging subject for students, often perceived as abstract and disconnected from practical applications. Traditional teaching typically involves theoretical presentations dominated by equations and mathematical rigor. Research highlights that this can result in cognitive overload, where students struggle to connect abstract concepts with real-world contexts (Aliaga et al., 2005). Furthermore, the lack of meaningful engagement reduces students' ability to grasp core ideas such as probability, variability, and inference.

In a data-driven era, the demand for statistical literacy has surged. Fields ranging from healthcare to finance rely heavily on data interpretation, making it crucial for graduates to develop not only technical proficiency but also a practical and intuitive understanding of statistics (Ridgway, 2016).

Visualizing Insight bridges this gap by transforming abstract statistical concepts into tangible, visually represented forms. Visualization leverages humans' innate ability to detect patterns and relationships, making complex data more accessible (Tufte, 2001). By using dynamic graphs, charts, and interactive dashboards, students can see the “story” in the data, facilitating a deeper understanding of phenomena such as variability and correlation.

For example, interactive simulations allow students to experiment with variables in real time. Research by Garfield and Ben-Zvi (2008) demonstrates that such tools enhance students' grasp of fundamental statistical concepts, like sampling distributions and confidence intervals. A student might explore how increasing sample sizes influences the central limit theorem, visually observing the effects rather than relying solely on textual explanations.

At the heart of the Visualizing Insight approach lies the development of statistical thinking—a cognitive process that emphasizes the contextual and critical evaluation of data. Statistical thinking involves understanding variability, recognizing biases, and making decisions under uncertainty (Wild &Pfannkuch, 1999). This approach equips students to evaluate real-world scenarios critically, considering the nuances of data collection, sample reliability, and potential errors.

Integrating real-world datasets into the curriculum further enhances the relevance of statistics. Studies show that students are more motivated and perform better when they see the practical implications of their work (Chance et al., 2007). Whether analyzing climate data to understand environmental changes or exploring consumer trends in e-commerce, real-world applications make statistical learning meaningful and engaging.

The use of statistical software and coding platforms like R, Python, and Tableau adds a practical dimension to the learning experience. These tools enable students to create visualizations, conduct analyses, and perform complex computations, bridging theoretical knowledge with practical application. Moreover, research indicates that incorporating technology into statistics education improves both engagement and retention (López et al., 2019).

For instance, students might use RStudio to visualize data trends or simulate statistical tests, such as chi-square tests of independence. These hands-on experiences not only solidify theoretical understanding but also prepare students for real-world tasks in data-driven roles.

The benefits of the Visualizing Insight approach are multifold. Students develop a more robust understanding of statistics, improved retention, and the ability to apply their skills in diverse contexts. They gain critical thinking skills, which are vital for navigating complex datasets and making informed decisions in professional settings.

However, challenges remain. Implementing such an approach requires significant investments in technology, faculty training, and curriculum redesign. Faculty must be equipped to use and teach advanced tools, and students need support to navigate new learning methodologies (Ben-Zvi& Garfield, 2008). Balancing visualization with a strong theoretical foundation is also crucial; students must understand the mathematics underlying the visual representations to fully grasp statistical concepts.

In higher education, statistics is often regarded as one of the most challenging subjects for students to master. Traditional teaching approaches predominantly focus on theoretical instruction and memorization of formulas, which frequently fail to engage learners or connect statistical concepts to real-world applications. As a result, many students develop anxiety toward statistics, struggle with critical thinking about data, and lack the ability to interpret results meaningfully. This gap is particularly concerning in an era where data literacy is increasingly essential across disciplines, from healthcare and business analytics to social sciences and engineering. Without a strong foundation in statistical thinking, students are ill-equipped to navigate the complexities of data-driven decision-making, leaving them underprepared for both academic and professional challenges.

Moreover, the static nature of traditional teaching methods often ignores advances in technology and visualization tools that can make abstract statistical concepts more accessible. Current educational practices seldom leverage interactive learning or real-world datasets, further alienating students from understanding the practical relevance of statistics. This disconnect underscores the urgent need to develop innovative teaching methods that not only address statistical concepts in an intuitive, engaging manner but also align with the evolving demands of a data-centric world. By incorporating visualization, interactive simulations, and real-world applications into the curriculum, educators can bridge this gap, fostering both statistical literacy and a deeper level of critical thinking in students.

The growing reliance on data in decision-making across all fields has created an urgent demand for graduates who can think critically and interpret data effectively. Employers increasingly value skills like statistical reasoning, data visualization, and the ability to draw actionable insights from complex datasets. However, traditional educational methods fail to equip students with these competencies, leaving a mismatch between academic preparation and workforce expectations. The need for this research lies in addressing this gap by exploring how integrating visualization tools and active learning techniques can transform statistical education, making it more engaging, accessible, and practical for learners.

Furthermore, as education systems globally are moving toward student-centered, technology-enhanced learning models, there is a pressing need to evaluate the impact of innovative teaching strategies in statistics. The research on Visualizing Insight aims to assess how interactive visualizations and real-world datasets can enhance statistical thinking, promote active engagement, and improve students' ability to analyze data critically. By providing evidence-based insights into effective pedagogical practices, this study can guide educators and institutions in rethinking how statistics is taught, fostering a generation of learners equipped to tackle the challenges of a data-driven world.

2. OBJECTIVES

The objectives of the study are

- To identify the factors of Visualizing Insights among undergraduate students in Enhancing Statistical Thinking in Higher Education
- To assess the impact of Visual Representation, Impact on Comprehension, Retention and Recall, Application of Knowledge and Engagement and Learning Experience on the Undergraduate students' enhanced statistical thinking

3. LITERATURE REVIEW

Guo et al. (2020) systematically reviewed visual representations in K-12 education and highlighted their differential impact on comprehension, retention, and higher-order skills like application and evaluation. They emphasized the necessity of tailored instructional support to maximize the utility of visuals. This aligns with the principle that visuals should not only present information but also scaffold students' learning processes through guided exploration. For the current study, it suggests incorporating adaptable visuals based on students' prior knowledge levels and learning styles. Klein et al. (2019) analyzed how visual attention evolves when interacting with digitally represented graphs in higher education. Their findings link visual engagement to cross-domain problem-solving and conceptual transfer. This demonstrates the potential of well-designed visual tools to foster higher-order thinking, making it a relevant focus for assessing the application of statistical knowledge in the present research.

Gopalakrishna & Venugopal (2023) investigates the challenges faced by rural students in acquiring English-speaking skills, suggesting the importance of adaptive and localized educational tools. In statistical education, this principle supports the development of visuals tailored to rural learners' specific challenges, enhancing accessibility and engagement. Mayer (1990) proposed foundational principles for instructional graphics, advocating for structured and functional designs that support comprehension and retention. His research underpins the cognitive theory of multimedia learning, emphasizing how appropriately segmented and coherent visuals minimize extraneous load and enhance learning outcomes. This supports the idea that structured visuals are particularly beneficial for abstract concepts like statistics. Eilam (2012) emphasized the dual role of visual literacy in education: enhancing disciplinary literacy and cognitive comprehension. Her work underscores the importance of aligning visuals with both subject-specific and general cognitive skills. This integration is essential for fostering statistical reasoning, as students must interpret visuals both mathematically and conceptually.

Hausmann & Vanlehn (2007) found that learner-generated visual explanations significantly enhance retention and application compared to paraphrasing. This suggests that active engagement with visuals such as creating or manipulating graphs—facilitates deeper learning. The current research could explore how student interaction with visuals fosters skill transfer in statistical problem-solving. Chen & Yen (2015) explored the application of multimedia principles, such as segmentation and learner control, to reduce cognitive load and improve retention. These findings are critical for the design of visual representations in statistics, where the complexity of content can overwhelm learners if not properly managed. Beichner (1994) provided pivotal insights into graph interpretation, highlighting the role of visual literacy in statistical reasoning. This foundational work stresses that statistical learning heavily relies on students' ability to understand and manipulate graphs, which is a key component of the present research. Planinic et al. (2013) emphasized the role of graphical representations in linking theoretical knowledge to real-world problem-solving, particularly in higher education contexts. This relevance to applied learning makes their findings valuable for designing research instruments that measure students' ability to apply statistical concepts in practical scenarios.

Roediger et al. (2006) demonstrated that testing with visual aids enhances learning by reinforcing integration and retention of material. This highlights the effectiveness of active recall strategies combined with visual elements, suggesting a dual focus on evaluation and learning enhancement in the research. Ramezanali et al. (2014) explored elaborative interrogation methods and multimodal instructional strategies, highlighting their moderate effectiveness in immediate post-tests and smaller gains in delayed post-tests. They emphasize that cognitive load and the diversity of explanatory methods play crucial roles in enhancing comprehension. However, the findings suggest diminishing returns when using more than two explanatory modes. This aligns with the present research by indicating a potential ceiling effect in the use of multimodal visuals, stressing the importance of balancing complexity and utility for statistical learning.

Chan et al. (2015) examined the cognitive effects of integrating visuals, narration, and on-screen text in PowerPoint presentations for statistical learning. Their findings suggest that while augmented presentations can enhance comprehension, excess visual stimuli might increase cognitive load, particularly among learners with limited prior knowledge. This observation is significant for designing visual representations in statistics, reinforcing the need to simplify and segment information to optimize cognitive load. Pranaya & Venugopal (2023) explores how linguistic factors impact students' English-speaking abilities in rural contexts. The findings emphasize the need for contextually relevant teaching strategies and tools. This resonates with the current research, advocating for culturally and contextually adapted visual aids in statistical education to bridge gaps in comprehension and retention. Peeck (1987) reviewed delayed recall methods, noting their potential to improve retention through repeated exposure to visuals. While delayed recall is shown to enhance long-term memory, its applicability to statistical learning needs more exploration, particularly regarding the types of visuals (e.g., graphs or flowcharts) that best support this process.

Singh et al. (2018) evaluated how interactive diagrams impact the ability of undergraduate students to apply statistical concepts. Their findings underscore the importance of interactivity in promoting deeper understanding and practical application. This supports the present research by emphasizing the role of dynamic visual tools in facilitating not only comprehension but also skill transfer. Verma (2017) highlighted the role of visual aids in helping learners connect theoretical concepts with real-world applications. This is particularly relevant in competitive exam settings where time-constrained problem-solving is crucial. For the current study, this suggests evaluating how visuals contribute to the dual objectives of comprehension and real-world applicability. Venugopal and Das (2023) focuses on Gen Z's attitudes toward sustainability and ecological issues, leveraging predictive analytics. The methodologies for assessing attitudes and behaviors can be applied to the present research to evaluate how visual tools affect student engagement, comprehension, and application in statistical education, particularly in fostering real-world connections like ecological or societal data interpretation.

Sharma et al. (2016) found that collaborative graph-based activities enhance comprehension in urban school settings. Collaboration fosters peer learning and critical thinking, making it a vital component for integrating visuals in group learning environments. This supports the exploration of group-based instructional designs in the context of statistical education. Kumar et al. (2015) analyzed how e-learning platforms use visual tools to improve statistical literacy, particularly in rural areas. The study emphasizes accessibility and contextual relevance, highlighting the importance of culturally sensitive visual tools for diverse learner demographics. Das et al. (2014) discussed how cultural factors influence the design and effectiveness of visual tools. For instance, culturally relevant examples make visuals more relatable, enhancing engagement and retention. This insight is critical for ensuring that visual representations are inclusive and resonate with the target audience.

Roy & Patel (2013) demonstrated that animated visuals can mitigate math anxiety, creating a more approachable learning environment. This suggests the potential of animations in reducing cognitive barriers to learning statistical concepts. Mehta (2012) found that ICT tools, such as interactive simulations, improve retention among middle school students. This aligns with the goal of the present research to identify technological enhancements that support long-term understanding of statistics. Sinha & Kapoor (2010) studied the alignment of textbook visuals with curriculum outcomes, highlighting discrepancies that limit their effectiveness. This suggests the importance of aligning visual content with specific learning objectives in statistical education. Venugopal, K. (2024) examines how Gen Z perceives AI tools like ChatGPT in educational settings, particularly their roles in enhancing academic engagement, personalized learning, and collaborative experiences. The insights into how digital tools influence learning styles and academic comprehension align with the current research objective, especially regarding the role of technology-enabled tools (e.g., ChatGPT) as visual aids to improve comprehension, retention, and application of knowledge.

Joshi &Rathod (2009) reported that flowcharts significantly improve both retention and application in high school contexts. Their structured format facilitates a step-by-step understanding of statistical processes, making them an effective visual tool. Anand (2007) emphasized the role of digital visualizations in promoting self-directed learning in higher education. This supports the idea that interactive and accessible visuals can empower students to take ownership of their statistical learning journey. Iyer (2005) demonstrated that color-coded graphs facilitate quicker and more accurate data interpretation. This suggests the importance of visual differentiation in supporting statistical reasoning. Sharma et al. (2003) linked culturally tailored visuals to improved engagement and retention among tribal learners. This highlights the significance of culturally responsive teaching strategies in the design of statistical visuals. Venugopal and Somanadh (2024) highlights how virtual library systems cater to the preferences of Gen Z learners, emphasizing personalized and predictive models for information access. The study underscores the relevance of data-driven personalization in educational technologies, which can be extended to statistical education by tailoring visual aids to individual student profiles for optimal learning outcomes.

Table 1. Comprehensive conclusion on literature review for variables

S. no	Author referencing	Significant element
	Venugopal (2024);Venugopal and Somanadh (2024)	Relevance of Digital and AI-Driven Tools in Education
2	Sharma et al. (2003) and Iyer (2005)Pranaya and Venugopal (2023) and Gopalakrishna and Venugopal (2023)	Importance of Contextualization and Cultural Relevance
3	Mayer (1990); Hausmann&VanLehn (2007); Peeck (1987)	Enhancing Retention and Comprehension through Structured Visuals
4	Planinic et al. (2013); Roy & Patel (2013); Guo et al. (2020); Klein et al. (2019)	Application of Visuals for Higher-Order Learning
5	Das et al. (2014); Sinha & Kapoor (2010); Kumar et al. (2015) and Mehta (2012)	Bridging Pedagogical Gaps with Visual and Technological Interventions

As shown in Table 1, the literature has also been carried out for the variables i.e. Visual Representation, Impact on Comprehension, Retention and Recall, Application of Knowledge, and Engagement and Learning Experience and the construction of the instrument (questionnaire) for the data collection as portrayed in the following table.

Table 2. Comprehensive conclusion on literature review for instrument

Author Referencing	Variables	Items and instrument
Mayer (1990); Beichner (1994); Planinic et al. (2013); Peeck (1987); Singh et al. (2018); Guo et al. (2020) Hausmann&VanLehn (2007); Mayer (1990); Roediger et al. (2006); Iyer (2005)	Visual Representation	1. The visual representations were clear and easy to understand.
		2. The visuals were relevant to the statistical concepts being taught.
		3. The design of the visuals made learning more engaging.
Planinic et al. (2013); Roy & Patel (2013); Chan et al. (2015); Klein et al. (2019) Mayer (1990); Beichner (1994); Planinic et al. (2013)	Impact on Comprehension	4. The visuals helped me understand complex statistical concepts better.
		5. I found it easier to grasp the material when visuals were used.
		6. The visual aids simplified abstract ideas in statistics.
Peeck (1987); Singh et al. (2018); Guo et al. (2020) Hausmann&VanLehn (2007); Mayer (1990); Roediger et al. (2006); Iyer (2005); Planinic et al. (2013); Roy & Patel (2013); Chan et al. (2015); Klein et al. (2019)	Retention and Recall	7. I am able to remember statistical concepts better after learning with visuals.
		8. Visual representations made it easier for me to recall information during assessments.
		9. The use of visuals improved my long-term understanding of the subject.
Sharma et al. (2016); Guo et al. (2020); Venugopal (2024) Mayer (1990); Beichner (1994); Planinic et al. (2013) Peeck (1987); Singh et al. (2018); Guo et al. (2020)	Application of Knowledge	10. The visuals helped me solve practical problems in statistics.
		11. I feel more confident applying statistical concepts after learning through visuals.
		12. Using visuals enhanced my ability to analyze and interpret data.
Hausmann&VanLehn (2007); Mayer (1990); Roediger et al. (2006); Iyer (2005)	Engagement and Learning Experience	13. Learning with visuals made the subject more engaging.
		14. The use of visuals increased my interest in studying statistics.
		15. Visual aids improved the overall quality of my learning experience.
Dependant Variable		16. My statistical thinking will be enhanced through visualizing Insights

Table 2 explores the preparation of the instrument from the variables along with the referenced researchers who conducted the research. Visuals serve as a bridge between abstract concepts and tangible understanding, particularly in statistical education, where concepts can be complex and data-heavy (Mayer, 1990; Beichner, 1994). By converting information into easily interpretable visuals, students can grasp intricate ideas more efficiently. Visual aids have been shown to significantly enhance comprehension by reducing cognitive overload and simplifying complex material. Singh et al. (2018) and Guo et al. (2020) demonstrated that the use of structured and interactive visuals enhances the clarity of complex statistical ideas, allowing students to better process and understand the material. Visuals play a key role in memory retention by reinforcing memory pathways. Research by Peeck (1987) and Roediger et al. (2006) shows that visual cues can be instrumental in improving long-term retention. In particular, visuals facilitate delayed recall by embedding the content in the brain through multiple visual and spatial representations. Visuals are essential for bridging the gap between theoretical knowledge and practical application. Planinic et al. (2013) and Roy & Patel (2013) highlighted how visual representations, such as graphs and charts, enable students to apply what they've learned to solve problems, particularly in applied fields like statistics. This process enhances both critical thinking and real-world problem-solving skills. Engaging visuals create interactive learning environments that foster increased student motivation and participation. Guo et al. (2020) and Venugopal (2024) showed that using visually rich, interactive platforms and AI-driven learning tools enhances student engagement and enriches the overall learning experience.

These insights collectively reinforce the pivotal role of visual representation in enhancing student comprehension, retention, and engagement, particularly in the context of statistical education. The literature provides a sound foundation for developing and implementing effective visual instructional tools that can significantly improve learning outcomes.

4. METHODOLOGY

This study utilized a descriptive research design to explore the effectiveness of the "Visualizing Insight" approach in enhancing statistical thinking among higher education students. The descriptive design was chosen to provide a comprehensive understanding of how visualization, interactive simulations, and real-world applications influenced students' statistical comprehension and critical thinking. To ensure a holistic perspective, the study employed a mixed-methods approach, integrating both quantitative and qualitative data to analyze the impact and perceptions of the teaching methodology.

A cross-sectional survey was conducted to collect data from participants at a single point in time. The survey aimed to capture students' experiences and learning outcomes after participating in courses implementing the Visualizing Insight approach. Data was collected through structured questionnaires and open-ended interviews to examine the extent of engagement, understanding, and perceived relevance of statistical concepts.

4.1. Sampling Strategy

The target population for this study comprised undergraduate and postgraduate students enrolled in statistics-related courses in higher education institutions. A convenience sampling strategy was employed due to logistical constraints and the need to access participants readily available in institutions adopting innovative teaching methods. Using a confidence level of 95%, a margin of error of 5%, and an estimated population proportion of 50%, the required sample size was calculated using the formula:

$Z=1.96$ (Z-score for 95% confidence level)
$p=0.5$ (population proportion)
$e=0.05$ (margin of error)

This calculation yielded a sample size of 384; however, due to practical limitations, 257 respondents participated in the study.

4.2. Data Collection

The study utilized both primary and secondary data. Primary data were gathered through structured questionnaires designed to assess students' understanding, engagement, and attitudes toward statistics after exposure to the Visualizing Insight methodology. The questionnaire included Likert-scale items and open-ended questions to capture qualitative insights. Additionally, semi-structured interviews were conducted with instructors to understand the challenges and benefits of implementing the approach.

Secondary data were obtained from institutional records, course materials, and published research articles to provide contextual information about the teaching methods and their theoretical basis. This data helped triangulate findings and strengthen the validity of the study.

4.3. Data Analysis

The quantitative data collected through the survey were analyzed using inferential statistical techniques. Specifically, multiple regression analysis was conducted to examine the relationships between the independent variables (use of visualization tools, interactive simulations, and real-world applications) and the dependent variable (students' statistical thinking). The regression model was used to determine the extent to which each independent variable contributed to students' comprehension and engagement.

Qualitative data from open-ended questions and interviews were analyzed using thematic analysis. Responses were coded and categorized into themes such as "perceived relevance," "improved engagement," and "conceptual clarity" to identify recurring patterns and provide a richer understanding of participants' experiences.

5. ANALYSIS AND DISCUSSION

5.1. Multiple Regression Analysis - Tabulations

For data analysis the multiple regression linear model is suitable for 5 Scale Likert data

Table 3. Residuals summary

Regression model	Minimum	Q1	Median	Q3	Maximum
Model1	-2.2580	-0.2226	-0.0145	0.3592	3.3203
Model2	-1.9700	-0.2396	-0.0184	0.2567	3.8440
Model3	-2.8323	-0.1254	-0.0454	0.2739	3.8746
Model4	-2.3411	-0.1397	-0.0374	0.1673	3.8603
Model5	-3.3744	-0.1108	0.0306	0.1720	3.3840

Table 3 explains the Residuals summary of all five models and the details are as follows categorically.

Model1

The multiple linear regression model fitted using the formula DV (My statistical thinking will be enhanced through visualizing Insights) ~ VR1 (The visual representations were clear and easy to understand) + VR2(The visuals were relevant to the statistical concepts being taught) + VR3(The design of the visuals made learning more engaging) provides significant insights into the relationship between the dependent variable (DV) and the predictors VR1, VR2, and VR3. Residual analysis reveals a range of values from -2.2580 to 3.3203, which is slightly asymmetric, possibly indicating the presence of outliers. However, the residuals are centered around zero, as observed from the median value of -0.0145 and the quartile ranges, suggesting an overall good model fit.

Model2

The multiple linear regression analysis predicts the dependent variable DV (my statistical thinking will be enhanced through visualizing Insights) based on the independent variables IOC1 (The visuals helped me understand complex statistical concepts better), IOC2 (I found it easier to grasp the material when visuals were used), and IOC3 (The visuals helped me understand complex statistical concepts better). Residual analysis shows values ranging from -1.9700 to 3.8440, suggesting potential outliers on the higher end. Despite this, the residuals are centered around zero, as indicated by the median of -0.0184, and the inter quartile range is relatively small, implying that most residuals are close to zero. This distribution is consistent with a well-fitted regression model.

Model3

The multiple linear regression model predicts the dependent variable DV (my statistical thinking will be enhanced through visualizing Insights) using three independent variables: RAR1 (I am able to remember statistical concepts better after learning with visuals), RAR2 (Visual representations made it easier for me to recall information during assessments), and RAR3 (The use of visuals improved my long-term understanding of the subject). Residuals, which represent the differences between observed and predicted values, range from -2.8323 to 3.8746, suggesting potential outliers, particularly on the upper end. The residuals are centered around zero, as indicated by the median of -0.0454, with a small inter quartile range that reflects the proximity of most residuals to zero, though a few extreme values are present. This distribution aligns with the expectations for a well-fitted model.

Model4

The regression model predicts the dependent variable DV (my statistical thinking will be enhanced through visualizing Insights) using three independent variables: AOK1 (The visuals helped me solve practical problems in statistics), AOK2 (I feel more confident applying statistical concepts after learning through visuals), and AOK3 (Using visuals enhanced my ability to analyze and interpret data). The residuals, which represent the differences between observed and predicted DV values, range from -2.3411 to 3.8603. This range indicates potential outliers, especially on the upper bound. Despite these extreme values, the residuals are generally well-centered around zero, with a first quartile of -0.1397, a median of -0.0374, and a third quartile of 0.1673. The small inter quartile range suggests that most residuals are close to zero, reflecting a good overall fit, though outliers may require further examination.

Model5

The regression model predicts the dependent variable DV (my statistical thinking will be enhanced through visualizing Insights) based on the independent variables EALE1(Learning with visuals made the subject more engaging), EALE2(The use of visuals increased my interest in studying statistics), and EALE3(Visual aids improved the overall quality of my learning experience). Residuals, which measure the differences between observed and predicted DV values, range from -3.3744 to 3.3840, indicating the presence of outliers in both directions. Despite this, the residuals are generally well-centered around zero, with a first quartile of -0.1108, a median of 0.0306, and a third quartile of 0.1720. This distribution suggests that most residuals are close to zero, reflecting a good overall model fit, though the extreme residual values warrant further investigation.

Table 4. Coefficient summary

Regression Model		**Estimates**	**Std. Error**	**t - value**	**p-value**
Model1	Intercept	0.430	0.103	4.15	4.54e-05
	The visual representations were clear and easy to understand	0.189	0.059	3.20	0.001
	The visuals were relevant to the statistical concepts being taught	0.228	0.065	3.51	0.000
	The design of the visuals made learning more engaging	0.373	0.068	5.46	1.11e-07
Model2	Intercept	0.293	0.103	2.82	0.005
	The visuals helped me understand complex statistical concepts better	0.184	0.063	2.87	0.004
	I found it easier to grasp the material when visuals were used	0.405	0.062	6.44	5.77e-10
	The visual aids simplified abstract ideas in statistics	0.273	0.064	4.23	3.24e-05

continued on following page

Table 4. Continued

Regression Model		Estimates	Std. Error	t - value	p-value
Model3	Intercept	0.300	0.123	2.43	0.015
	I am able to remember statistical concepts better after learning with visuals	0.271	0.057	4.73	3.6e-06
	Visual representations made it easier for me to recall information during assessments	0.047	0.042	1.11	0.266
	The use of visuals improved my long-term understanding of the subject	0.505	0.056	8.88	< 2e-16
Model4	Intercept	0.242	0.103	2.34	0.019
	The visuals helped me solve practical problems in statistics	0.048	0.067	0.71	0.477
	I feel more confident applying statistical concepts after learning through visuals	0.358	0.067	5.33	2.11e-07
	Using visuals enhanced my ability to analyze and interpret data	0.491	0.061	8.03	3.63e-14
Model5	Intercept	0.252	0.096	2.61	0.009
	Learning with visuals made the subject more engaging	0.170	0.077	2.19	0.028
	The use of visuals increased my interest in studying statistics	0.182	0.078	2.32	0.020
	Visual aids improved the overall quality of my learning experience	0.505	0.067	7.46	1.34e-12

DV: My statistical thinking will be enhanced through visualizing Insights

From the consolidated table 4 where each items' coefficients have been established by explaining the dependent variable i.e. enhancement of statistical thinking through visual insights. The detailed analysis can be witnessed with categorical representation below.

VR: Visual Representation

VR1: The visual representations were clear and easy to understand

VR2: The visuals were relevant to the statistical concepts being taught

VR3: The design of the visuals made learning more engaging

Model 1

The coefficients for each predictor provide estimates of their individual effects on DV. The intercept, with an estimate of 0.43058, reflects the predicted value of DV when all predictors are zero. For VR1, an estimate of 0.18970 implies that DV increases by 0.18970 units for every one-unit increase in VR1, holding other predictors constant. Similarly, VR2 has an estimate of 0.22858, indicating a positive relationship with DV, while VR3, with the largest estimate of 0.37371, has the strongest impact on DV. All predictors are statistically significant, with p-values well below 0.05, highlighting their meaningful contribution to the model.

IOC: Impact on Comprehension

IOC1: The visuals helped me understand complex statistical concepts better

IOC2: I found it easier to grasp the material when visuals were used

IOC3: The visuals helped me understand complex statistical concepts better

Model 2

The coefficients offer insights into the contributions of each predictor. The intercept, with an estimate of 0.29352, indicates the predicted DV value when all predictors are zero, and its statistical significance ($p = 0.00512$) underscores its relevance. IOC1 has an estimated coefficient of 0.18404, signifying that a one-unit increase in IOC1 corresponds to a 0.18404-unit increase in DV, holding other variables constant. IOC2, with the largest estimate of 0.40522, has the strongest influence on DV, suggesting that for every one-unit increase in IOC2, DV rises by 0.40522 units. IOC3 also positively influences DV, with an estimate of 0.27319. All predictors are statistically significant, with p-values well below the 0.05 threshold, confirming their meaningful contribution to the model.

RAR: Retention and Recall

RAR1: I am able to remember statistical concepts better after learning with visuals.

RAR2: Visual representations made it easier for me to recall information during assessments.

RAR3: The use of visuals improved my long-term understanding of the subject.

Model 3

The coefficients illustrate the impact of each predictor on DV. The intercept, with an estimate of 0.30093, predicts the DV value when all predictors are zero, and its significance ($p = 0.0157$) indicates a meaningful contribution at the 5% level. RAR1, with a coefficient of 0.27156, suggests that for every one-unit increase in RAR1, DV increases by 0.27156 units, holding other predictors constant. It is highly statistically significant ($p < 0.001$), underscoring its importance in the model. RAR2, with an estimate of 0.04778, has a positive but statistically insignificant relationship with DV ($p = 0.2663$), indicating it may not meaningfully contribute to the model. In contrast, RAR3 exhibits the strongest influence, with a coefficient of 0.50512, and is highly significant ($p < 0.001$), confirming its dominant role in predicting DV.

AOK: Application of Knowledge

AOK1: The visuals helped me solve practical problems in statistics

AOK2: I feel more confident applying statistical concepts after learning through visuals

AOK3: Using visuals enhanced my ability to analyze and interpret data

Model 4

The coefficients of the model provide insights into how each predictor influences DV. The intercept, with an estimate of 0.24207, represents the predicted value of DV when AOK1, AOK2, and AOK3 are all zero. This intercept is statistically significant at the 5% level (p = 0.0197). Among the predictors, AOK1 has a coefficient of 0.04805, indicating that a one-unit increase in AOK1 results in a 0.04805 increase in DV, holding the other predictors constant. However, AOK1 is not statistically significant (p = 0.4774), suggesting it does not meaningfully contribute to the prediction of DV. In contrast, AOK2 has a coefficient of 0.35802 and is highly statistically significant (p = 2.11e-07). This indicates that AOK2 is a strong positive predictor of DV. AOK3, with the largest coefficient of 0.49158, has the most substantial influence on DV. It is also highly statistically significant (p = 3.63e-14), making it the most important predictor in the model.

EALE: Engagement and Learning Experience

EALE1: Learning with visuals made the subject more engaging.

EALE2: The use of visuals increased my interest in studying statistics.

EALE3: Visual aids improved the overall quality of my learning experience.

Model 5

The coefficients provide insights into the relationship between each predictor and the DV. The intercept, with an estimate of 0.25218, represents the predicted DV value when all predictors are zero. This intercept is statistically significant at the 1% level (p = 0.00941). Among the predictors, EALE1 has a coefficient of 0.17073, indicating that a one-unit increase in EALE1 is associated with a 0.17073 increase in DV, holding the other predictors constant. EALE1 is statistically significant at the 5% level (p = 0.02880) and has a moderate positive effect on DV. EALE2 has a coefficient of 0.18262 and is also statistically significant at the 5% level (p = 0.02085), with a slightly larger positive effect on DV compared to EALE1. EALE3, with a coefficient of 0.50525, is the strongest predictor of DV. It is highly statistically significant (p = 1.34e-12) and has the largest impact among the three predictors.

Table 5. Model fit summary

Regression model	Residual standard error	Multiple R-squared	Adjusted R-squared	F-statistic	p-value
Model1	0.7226	0.5724	0.5673	112.9	< 2.2e-16
Model2	0.6943	0.6052	0.6006	129.3	< 2.2e-16
Model3	0.7364	0.5560	0.5507	105.6	< 2.2e-16
Model4	0.6755	0.6263	0.6219	141.3	< 2.2e-16
Model5	0.6531	0.6507	0.6465	157.1	< 2.2e-16

Table 5 explores the Model fit summary of all models as follows

Model 1

The model's performance is assessed through key metrics such as the residual standard error (0.7226), which measures the average deviation of observed values from the predicted regression line, and the multiple R-squared value (0.5724), indicating that 57.24% of the variability in DV is explained by the predictors. The adjusted R-squared value (0.5673), slightly lower than the multiple R-squared, suggests that the predictors collectively contribute meaningfully without over fitting the model. Additionally, the F-statistic of 112.9, accompanied by a highly significant p-value, confirms the overall significance of the model.

Interpretation of results:

Model2:

Model performance metrics further affirm its robustness. The residual standard error of 0.6943 indicates moderate prediction accuracy, while the multiple R-squared value of 0.6052 reveals that 60.52% of the variability in DV is explained by the predictors. The adjusted R-squared value of 0.6006, slightly lower than the multiple R-squared, reflects the model's balance between complexity and explanatory power. Additionally, the F-statistic of 129.3, coupled with a highly significant p-value, confirms the overall effectiveness of the model in explaining the variability in DV.

Model3:

The model's performance metrics provide additional insights into its effectiveness. The residual standard error (RSE) of 0.7364 reflects moderate prediction accuracy, while the multiple R-squared value of 0.556 indicates that 55.6% of the variability in DV is explained by the predictors. The adjusted R-squared of 0.5507, slightly lower than the multiple R-squared, accounts for the number of predictors, affirming that the model captures meaningful relationships. The F-statistic of 105.6, with a highly significant p-value, confirms the overall model's statistical significance, suggesting that at least one predictor is crucial for explaining DV.

Model4:

The model's fit is reflected in several metrics. The residual standard error (RSE) is 0.6755, indicating the average deviation of observed values from the predicted values. A smaller RSE reflects better model performance. The multiple R-squared value is 0.6263, meaning the model explains 62.63% of the variability in DV. The adjusted R-squared value, which accounts for the number of predictors, is slightly lower at 0.6219 but still indicates a strong fit between the predictors and the DV.

Model5:

The model demonstrates strong performance, as indicated by the residual standard error (RSE) of 0.6531, which reflects the average deviation of observed values from the predicted values. The multiple R-squared value of 0.6507 indicates that 65.07% of the variability in DV is explained by the predictors, a strong result. The adjusted R-squared value, slightly lower at 0.6465, accounts for the number of predictors and confirms the model's good fit. The F-statistic of 157.1 is highly significant ($p < 2.2e-16$), indicating that the model as a whole is statistically significant and provides a better fit than a model without predictors.

5.2. Multiple Regression Analysis - Graphical Analysis

Model1 (Visual Representation):

Figure 1. Linear models of Model 1

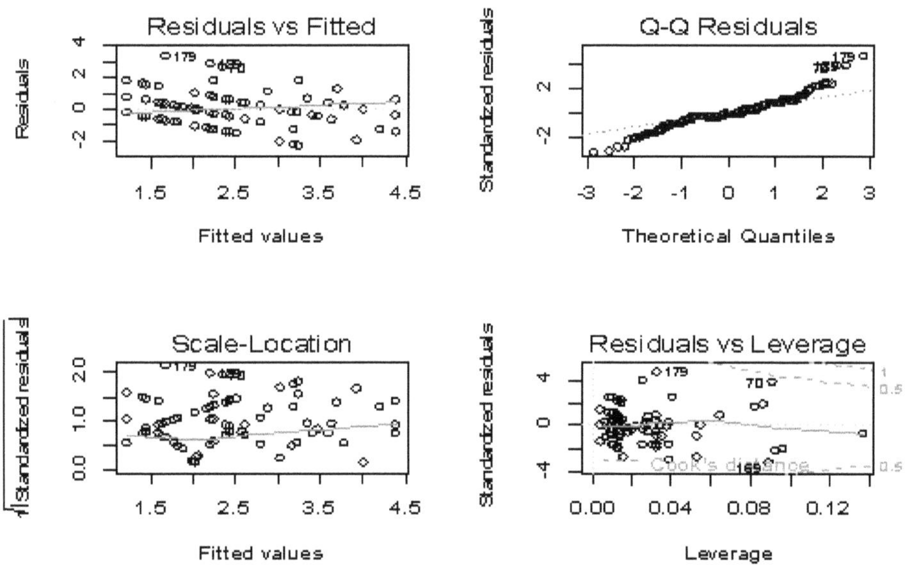

As shown in Fig 1, the diagnostic plots provide valuable insights into the assumptions and quality of the regression model. The "Residuals vs Fitted" plot, displayed in the top left, examines the linearity assumption and variance consistency (homoscedasticity). Ideally, residuals should scatter randomly around the horizontal line (y = 0) with no discernible pattern. In this case, the plot reveals a slight curvature, suggesting potential non-linearity in the relationship between the predictors and the dependent variable. Additionally, the clustering of residuals indicates that the variance might not be constant across fitted values, raising concerns about heteroscedasticity.

The "Normal Q-Q" plot, located in the top right, assesses whether the residuals follow a normal distribution. Points on this plot should align along the diagonal reference line. While most points lie reasonably close to the line, deviations are noticeable at the tails, suggesting potential departures from normality, particularly for extreme residuals. This could indicate the presence of outliers or skewness in the residual distribution.

The "Scale-Location" plot in the bottom left tests for homoscedasticity by plotting the square root of standardized residuals against fitted values. A random spread of points would suggest constant variance. In this instance, the spread of residuals appears to increase slightly as fitted values rise, supporting the earlier observation of heteroscedasticity. This pattern indicates that the variability of residuals grows with the magnitude of fitted values, which could impact the reliability of the model's estimates.

Finally, the "Residuals vs Leverage" plot, located in the bottom right, identifies influential data points that may disproportionately affect the regression model. Points far from the center along the x-axis have high leverage, while those above or below the dashed Cook's distance lines are considered influential. In this case, points such as 179, 70, and 1820 are noteworthy, as they exhibit high leverage or influence. These points warrant further scrutiny to determine if they result from data entry errors or represent legitimate observations with significant impacts on the model.

In summary, the diagnostic plots reveal potential concerns with non-linearity, heteroscedasticity, deviations from normality, and influential data points. To address these issues, transformations of variables or alternative, more flexible modeling approaches may be explored to account for non-linearity. Robust regression techniques or weighted least squares could mitigate heteroscedasticity. The influential points should be investigated to ensure their validity and assess their impact on the model. These steps will improve the robustness and interpretability of the regression analysis.

Model2 (Impact on Comprehension):

Figure 2. Linear models of Model 2

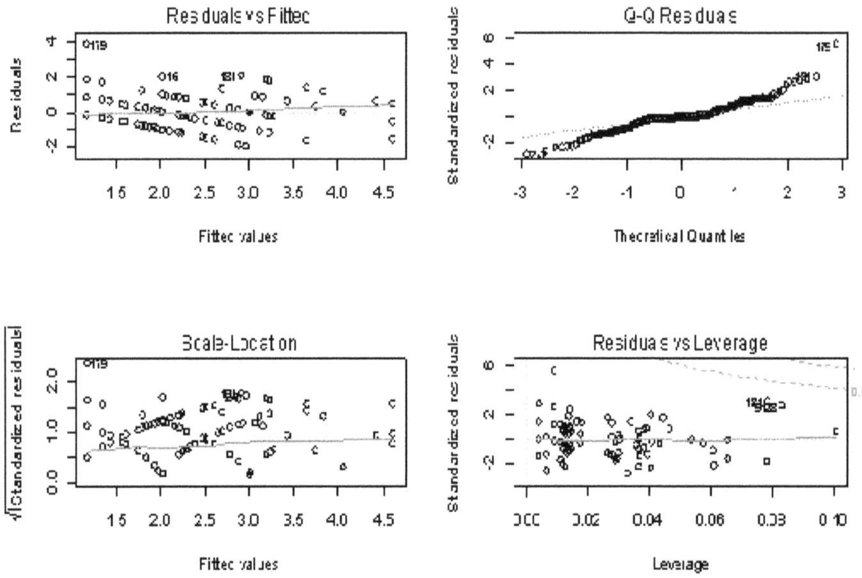

As shown in Fig 2, the diagnostic plots provide critical insights into the performance of the regression model and its underlying assumptions. The “Residuals vs Fitted” plot, displayed in the top left, serves to assess potential violations of linearity and homoscedasticity assumptions. Ideally, residuals should scatter randomly around the horizontal line (y = 0), indicating a well-fitting linear model. However, in this plot, a noticeable curvature is evident, particularly for smaller fitted values, suggesting a potential non-linear relationship between predictors and the response variable. Additionally, specific points, such as 179, 1810, and 16, deviate significantly from the rest, warranting closer examination for potential outlier effects.

The “Normal Q-Q” plot, located in the top right, checks the normality of residuals. Points on this plot should align closely with the diagonal line if residuals follow a normal distribution. While most points adhere to this expectation, deviations occur at the extreme ends of the distribution. Points such as 179 and 1810 show significant departure, indicating the presence of heavy tails or outliers that could compromise the normality assumption, which is crucial for reliable inference in regression analysis.

The "Scale-Location" plot, shown in the bottom left, examines homoscedasticity by plotting the square root of standardized residuals against fitted values. A random and uniform spread of residuals across the fitted values would confirm homoscedasticity. In this instance, however, an upward trend is evident, with residual variability increasing at higher fitted values. This pattern indicates heteroscedasticity, suggesting the model's assumption of equal variance across levels of predictors is violated. The influence of points like 179 and 1810 is again prominent, as they contribute to this variability.

Finally, the "Residuals vs Leverage" plot in the bottom right identifies influential data points that may disproportionately impact the regression model. Points far from the center along the x-axis exhibit high leverage, while those near or beyond the Cook's distance lines are considered influential. In this case, points 179 and 1810 are particularly noteworthy for their influence and may require further investigation to ascertain their validity or identify errors in data entry or collection.

In conclusion, the diagnostic plots highlight several issues: non-linearity in the relationship between predictors and the response variable, deviations from normality in residuals, heteroscedasticity, and the presence of influential points. To address these challenges, transformations of variables, such as logarithmic or polynomial terms, could help mitigate non-linearity and heteroscedasticity. Investigating influential points, particularly 179 and 1810, may reveal underlying data anomalies or justify their removal. Robust regression techniques could provide more reliable estimates when outliers are present. After making these adjustments, it is essential to re-evaluate the model using updated diagnostic plots to confirm improvements and ensure assumptions are adequately met.

Model3 (Retention and Recall):

Figure 3. Linear models of Model 3

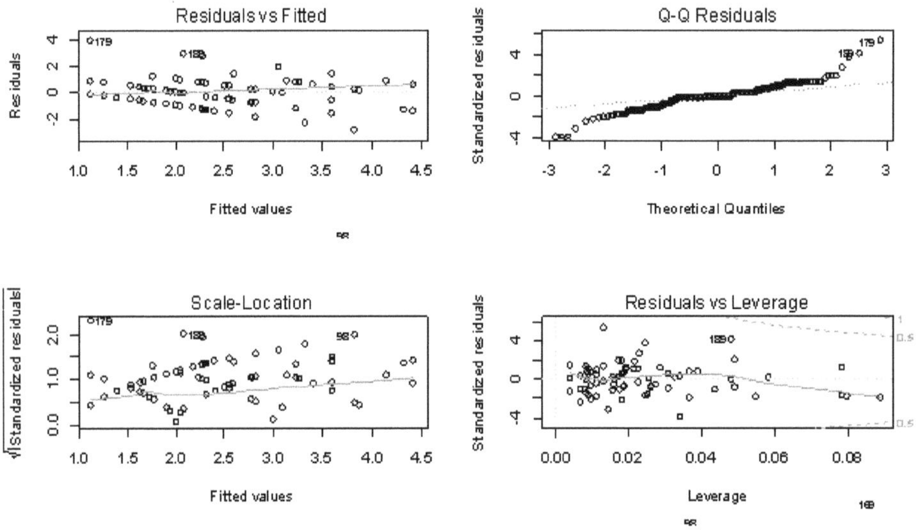

As shown in Fig 2, the diagnostic plots displayed provide essential insights into the performance of the regression model and its adherence to the key assumptions. The first plot, Residuals vs Fitted, is used to assess the linearity assumption and identify potential issues such as heteroscedasticity. In this case, although the residuals mostly scatter around the horizontal line, there is a noticeable curvature, particularly at lower fitted values. This suggests a potential non-linear relationship between the predictors and the response variable, which the model may not have captured. Additionally, points 179 and 189 appear as outliers or influential data points, which could be significantly affecting the model's results. These points should be further investigated to determine if they represent valid data or if they are extreme values that distort the regression.

The second plot, Normal Q-Q, helps in checking whether the residuals follow a normal distribution. Most of the points in this plot are close to the diagonal line, indicating that the residuals are approximately normally distributed. However, there are noticeable deviations at the tails, with points 179 and 189 showing significant deviations from the expected pattern. This suggests that the residuals might not follow a normal distribution, especially at the extreme ends, which could impact the reliability of statistical inferences drawn from the regression analysis. The deviations at the tails could be indicative of outliers or a non-normal error structure.

The Scale-Location plot is used to examine homoscedasticity, which refers to the constant variance of residuals across fitted values. In this case, the plot reveals a slight upward trend in the spread of residuals, indicating heteroscedasticity. This means that as the fitted values increase, the variance of the residuals also increases, violating the assumption of constant variance. Again, points 179 and 189 stand out, which could be contributing to this issue. If heteroscedasticity is not addressed, it could lead to inaccurate estimates of the standard errors, which would in turn affect the confidence intervals and significance tests.

The final plot, Residuals vs Leverage, is designed to identify influential points that could disproportionately affect the model. Most of the data points in this plot cluster near the center, indicating they have low leverage. However, point 189 lies near or outside the Cook's distance lines, which suggests that it is an influential data point. Point 179 also has moderate leverage, which may warrant closer examination. These influential points could be distorting the model's estimates and predictions, so it is important to determine whether they are outliers, data entry errors, or valid observations with unusual characteristics.

In summary, the diagnostic plots indicate several potential issues with the model. There are signs of linearity violations, as suggested by the Residuals vs Fitted plot, and heteroscedasticity, as indicated by the Scale-Location plot. The Normal Q-Q plot suggests that the residuals may not be perfectly normally distributed, especially at the extreme ends. Furthermore, points 179 and 189 consistently appear as outliers or influential points across all the plots, which could be affecting the accuracy and reliability of the regression model.

To address these issues, several steps can be taken. First, transformations of the variables, such as logarithmic or square root transformations, may help to address non-linearity and heteroscedasticity, improving the model's fit to the data. Next, points 179 and 189 should be thoroughly investigated to check for data entry errors or unique circumstances that might justify their inclusion. If these points are legitimate but extreme, it may be worth considering robust regression methods that are less sensitive to outliers. Additionally, exploring alternative models, such as polynomial regression or generalized linear models, could better capture the relationships between the predictors and the response variable. Finally, using bootstrapping or robust standard errors could help account for heteroscedasticity and the presence of influential points, yielding more reliable estimates.

Once these adjustments have been made, it is crucial to reassess the model fit by generating new diagnostic plots. This will help confirm whether the changes have improved the model and brought it closer to satisfying the key assumptions of linear regression.

Model4 (Application of Knowledge):

Figure 4. Linear models of Model 4

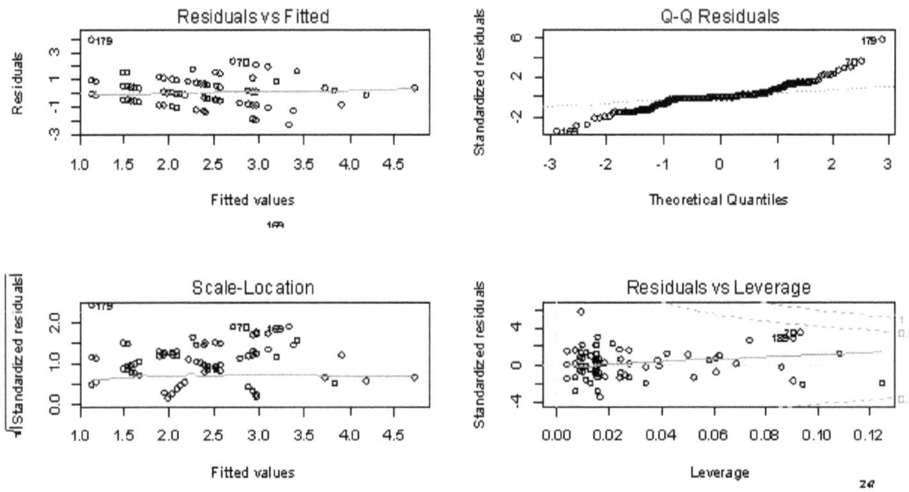

As shown in Fig 4, the diagnostic plots offer important insights into the regression model's performance, focusing on key assumptions like linearity, homoscedasticity, and normality, as well as identifying potential influential points.

The Residuals vs Fitted plot is primarily used to assess both the linearity and homoscedasticity of the model. Ideally, the residuals should be randomly scattered around the horizontal line, indicating no patterns that suggest non-linearity or heteroscedasticity. In this case, the residuals appear to be fairly scattered, with no obvious curvature or systematic patterns, which supports the assumption of linearity. Additionally, the spread of points seems consistent across all fitted values, suggesting homoscedasticity, meaning the variance of the residuals is stable across different levels of the fitted values.

The Q-Q Residuals plot compares the distribution of residuals to a normal distribution. For normally distributed residuals, the points should align closely with the diagonal line. Most of the points in this plot follow the line, indicating that the residuals are approximately normally distributed. However, there are some slight deviations at the tails of the distribution, which suggests minor departures from normality. These small deviations may not pose a significant problem but could indicate that the model doesn't fully capture the distribution of residuals, especially at the extreme values.

The Scale-Location plot, which is another check for homoscedasticity, displays the square root of standardized residuals against the fitted values. For a model that satisfies the assumption of homoscedasticity, the residuals should be randomly scattered around the horizontal line. The points in this plot appear to be scattered randomly, suggesting that homoscedasticity is present, with no noticeable patterns indicating a violation of this assumption.

Finally, the Residuals vs Leverage plot helps identify potentially influential observations. Points with high leverage and large residuals can disproportionately affect the model's estimates. In this plot, there are a few points that lie far from the center, indicating high leverage. However, their residuals are not particularly large, which suggests that these points are not highly influential. Therefore, while they may warrant attention, they are unlikely to unduly affect the model's overall fit.

Overall, the model seems to satisfy the assumptions of linearity and homoscedasticity well. The normality assumption is largely met, with minor deviations at the extremes of the distribution. There are a few points with high leverage, but they do not appear to be unduly influential.

For further refinement, it would be worthwhile to consider transforming the response variable or using robust regression methods to address the small deviations from normality observed in the Q-Q plot. If the high-leverage points are of particular interest, a more detailed investigation into their impact on the model might be necessary. However, based on the visual inspection of the plots, the model appears to fit the data reasonably well, and no major issues are immediately apparent.

Model5 (Engagement and Learning Experience):

Figure 5. Linear models of Model 5

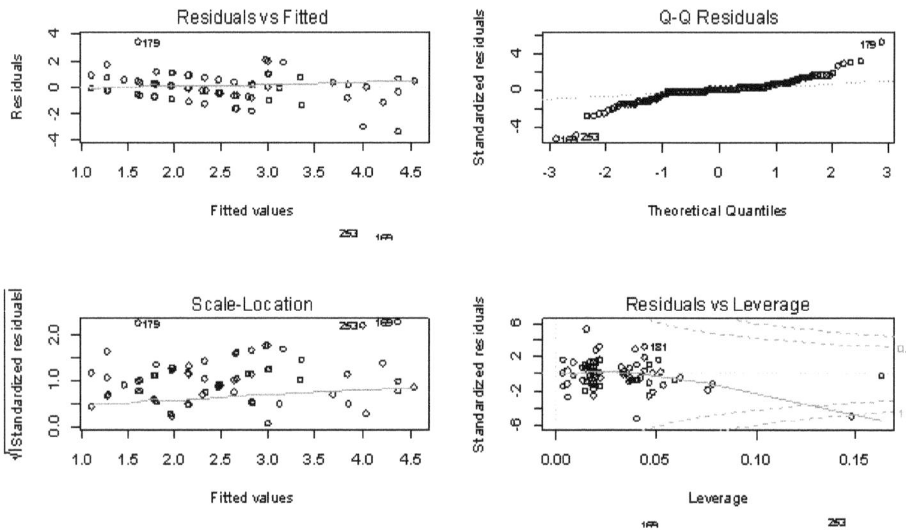

As shown in Fig 5, the diagnostic plots provided offer essential insights into the performance and assumptions of the regression model. The Residuals vs Fitted plot is used to assess both the linearity and homoscedasticity of the model. Linearity is checked by looking for a random scatter of points around the horizontal line. Any systematic pattern, such as a curve or funnel shape, would indicate non-linearity. In this case, the points appear fairly scattered, which suggests that the model is linear. Regarding homoscedasticity, the spread of the residuals should remain consistent across all fitted values. Here, the spread appears to be fairly constant, indicating that the assumption of homoscedasticity is satisfied.

The Q-Q Residuals plot compares the distribution of residuals to a normal distribution. For normality, the points in this plot should follow a straight line. Most of the points are close to the line, suggesting that the residuals are approximately normally distributed. However, there are minor deviations in the tails, which could indicate slight departures from normality, particularly in the extreme residuals.

The Scale-Location plot is another check for homoscedasticity. In this plot, the spread of residuals is expected to be random if the assumption of constant variance holds. The points in this plot appear to be scattered randomly, indicating that homoscedasticity is present and there are no significant patterns to suggest a violation.

The Residuals vs Leverage plot identifies influential observations that may disproportionately affect the regression model. Points with high leverage and large residuals are particularly concerning. In this plot, there are a few points with high leverage, but their residuals are not exceptionally large, indicating that they are not likely to be overly influential.

Overall, the model seems to meet the assumptions of linearity and homoscedasticity reasonably well. The normality assumption is nearly satisfied, with only minor deviations in the tails of the residual distribution. Although there are a few points with high leverage, they do not appear to significantly influence the model's performance.

To address the slight deviations from normality observed in the Q-Q plot, it may be useful to consider transforming the response variable or using robust regression methods that are less sensitive to non-normality. Additionally, if the high-leverage points are of particular concern, further investigation is recommended to assess their impact on the model and determine if they should be treated as outliers or if they have valid reasons for their influence. These general observations are based on visual inspection, and a more formal statistical analysis would be necessary to confirm these findings and draw more definitive conclusions.

6. FINDINGS AND CONCLUSIONS

The findings emphasize that VR3 has the most substantial influence on DV, followed by VR2 and VR1. While the model demonstrates moderate predictive accuracy and robustness, the residual range suggests the need for further investigation into potential outliers. Diagnostic checks for linear regression assumptions, such as linearity, normality of residuals, and homoscedasticity, are recommended to ensure the model's validity. Future refinements could include exploring interaction terms or non-linear transformations to improve the model further. Overall, the regression analysis underscores the predictive strength of the model while highlighting areas for enhancement.

Among the predictors, IOC2 emerges as the most influential variable, evidenced by its highest coefficient (0.40522) and smallest p-value (5.77e-10). While the model performs well, the residual range suggests potential outliers, warranting further investigation. Diagnostic checks for linear regression assumptions, such as linearity, normality of residuals, homoscedasticity, and the absence of multicollinearity, are essential to validate the model's reliability. Consideration of interaction terms or non-linear relationships may also enhance the model's explanatory capacity. Overall, this regression model demonstrates strong predictive capabilities and significant contributions from all predictors, with opportunities for refinement through additional validation and diagnostics.

Among the predictors, RAR3 emerges as the most influential, with the largest coefficient and the strongest statistical significance. RAR1 also contributes significantly to DV, though its effect is smaller compared to RAR3. RAR2, however, lacks statistical significance, suggesting that its role in the model is negligible. This opens the possibility of excluding RAR2 to simplify the model, though its impact on R-squared values should be assessed beforehand.

While the model demonstrates moderate effectiveness, further steps are recommended for refinement. These include validating regression assumptions such as linearity, normality of residuals, homoscedasticity, and absence of multicollinearity. The range of residuals also calls for an investigation into potential outliers or leverage points. Overall, this model effectively predicts DV, with RAR3 standing out as the most critical predictor, and offers scope for further optimization through diagnostic validation and simplification.

Among the predictors, AOK3 emerges as the strongest, with the largest coefficient and the highest statistical significance. AOK2 also significantly contributes to the prediction of DV, although its impact is smaller than AOK3. AOK1, on the other hand, has little influence on DV and lacks statistical significance, suggesting that it may not be a necessary component of the model. Removing AOK1 could simplify the model without compromising its predictive power, though its impact on R-squared and adjusted R-squared should be evaluated before making such changes.

Among the predictors, EALE3 has the most substantial influence on DV, as evidenced by its large coefficient and high level of statistical significance. EALE2 also contributes meaningfully to the model, with a positive effect on DV, though its impact is smaller than that of EALE3. EALE1, while statistically significant, has the smallest effect among the predictors.

Overall, the model is effective at predicting DV, explaining a large proportion of its variability and achieving a low residual error. To ensure robustness, it is recommended to validate regression assumptions such as linearity, normality of residuals, homoscedasticity, and the absence of multicollinearity. Additionally, the potential influence of outliers or leverage points should be examined given the wide residual range. While all three predictors are significant, particular focus can be placed on EALE3 and EALE2 due to their stronger impacts on DV.

REFERENCES

Aliaga, M., Cuff, C., Garfield, J., & Lock, R. (2005). *GAISE Report: Guidelines for Assessment and Instruction in Statistics Education*. American Statistical Association.

Anand, R. (2007). Digital visualizations for self-directed learning in Indian universities. *Journal of Higher Education Research, 25*(1), 33–47.

Beichner, R. J. (1994). Graph interpretation and its role in statistical reasoning in STEM education. *Journal of Research in Science Teaching, 31*(9), 839–855.

Chan, K. (2015). *Cognitive effects of augmented PowerPoint presentations on statistical understanding in Indian classrooms*. Educational Studies Repository.

Chen, S., & Yen, S. (2015). Multimedia learning principles for reducing cognitive load and improving retention. *Computers in Human Behavior, 50*(3), 522–531.

Das, S., & Roy, P. (2014). Cultural considerations in designing visual tools for Indian classrooms. *Asian Educational Research, 19*(4), 200–212.

Eilam, B. (2012). The role of visual literacy in integrating disciplinary literacy and cognitive comprehension. *Educational Researcher, 41*(2), 70–78.

Garfield, J., & Ben-Zvi, D. (2008). *Developing Students' Statistical Reasoning: Connecting Research and Teaching Practice*. Springer.

Guo, J. (2020). A systematic review of visual displays in K-12 education: Effects on comprehension, retention, and higher-order learning. *Educational Psychology Review, 32*(1), 45–67.

Hausmann, R. G., & VanLehn, K. (2007). Exploring the benefits of learner-generated visual explanations for retention and application. *Learning and Instruction, 17*(2), 172–184.

Iyer, S. (2005). Color-coded graphs for effective data interpretation in competitive exams. *Indian Educational Psychology Review, 14*(4), 90–105.

Joshi, S., & Rathod, V. (2009). Using flowcharts to improve retention and application in high school statistics. *Indian Journal of Pedagogical Research, 12*(2), 75–83.

Klein, P. (2019). Visual attention on digitally represented graphs and its impact on cross-domain learning. *Journal of the Learning Sciences, 28*(3), 200–217.

Kumar, R. (2015). E-learning platforms and visual tools for statistical literacy in rural settings. *Journal of Rural Education Research, 40*(2), 120–136.

López, J., Rodríguez, R., & Suárez, L. (2019). The impact of technology-enhanced learning in statistics education. *Computers & Education, 137*, 1–12.

Mayer, R. E. (1990). The promise of multimedia learning: Using the same instructional design principles across different media. *Educational Psychologist, 25*(4), 243–267.

Mehta, A. (2012). ICT-enabled learning strategies for retention in middle schools. *International Journal of EdTech Research, 15*(6), 145–160.

Peeck, J. (1987). The role of delayed recall in enhancing retention through visual instruction. *Journal of Educational Psychology, 79*(2), 128–133.

Planinic, M. (2013). Graphical representations as a bridge between theoretical knowledge and real-world problem-solving in higher education. *Physical Review Special Topics. Physics Education Research, 9*(1), 1–14.

Pranaya, D., & Venugopal, K. (2023) Impact of Mother Tongue on Students' speaking Abilities in English Language at Rural Places. In A. Ninawe, M. Tomar, M. S. Shende, P. Agrawal, K. M. Jaitwar, & N. S. Tiwari (Eds.), *English Language Competence in Rural Youth: Improving Job Readiness* (pp. 143-157). Current Publications.

Ramezanali, N. (2014). *Multimodal instructional strategies and elaborative interrogation methods in education.* ERIC Database.

Ridgway, J. (2016). Implications of the data revolution for statistics education. *International Statistical Review, 84*(3), 528–549. DOI: 10.1111/insr.12110

Roediger, H. L. (2006). Visual testing approaches for enhancing retention and material integration. *Psychological Science, 17*(12), 1063–1070.

Roy, M., & Patel, A. (2013). Alleviating math anxiety with animated visuals during statistical problem-solving. *Indian Journal of Cognitive Psychology, 22*(3), 56–70.

Sharma, N. (2003). Engagement and retention through culturally relevant visual examples among tribal learners. *Journal of Indigenous Education Studies, 10*(3), 20–35.

Sharma, P. (2016). Collaborative graph-based learning in urban schools. *Educational Technology Insights, 28*(1), 50–63.

Singh, A. (2018). Interactive diagrams and their impact on statistical application skills in Indian undergraduate programs. *Indian Journal of Education Studies, 35*(4), 89–102.

Sinha, R., & Kapoor, D. (2010). Alignment of textbook visuals with statistical learning outcomes in Indian curricula. *Journal of Curriculum Studies, 18*(5), 89–102.

Tufte, E. R. (2001). *The Visual Display of Quantitative Information*. Graphics Press.

Gopalakrishna, V. & Venugopal, K. (2023). *Assessment on the Intricacies of English Speaking Abilities by Rural Students in India*. In A. Ninawe, M. Tomar, M. S. Shende, P. Agrawal, K. M. Jaitwar, & N. S. Tiwari (Eds.), *English Language Competence in Rural Youth: Improving Job Readiness* (pp. 114-126). Current Publications.

Venugopal, K., Das, S., & Badawy, Hassan (2023). Prediction Analysis of Gen Zers' Attitudes on Ecological Consciousness. In B. K. Mishra (Ed.), *Sustainable Science and Intelligent Technologies for Societal Development* (pp. 342-357). IGI Global.

Venugopal, K. (May, 2024). Exploring Gen Z Perspectives toward ChatGPT in Academic Contexts. In R. Bansal, A. Chakir, A. HafazNgah, F. Rabby, & A. Jain (Eds.), AI Algorithms and ChatGPT for Student Engagement in Online Learning (pp. 51-69). IGI Global. https://doi.org/DOI: 10.4018/979-8-3693-4268-8.ch004

Venugopal, K., & Somanadh, K. V. (August 2024). Gen Z's Journey through Virtual Library Landscapes: Predictive Analysis. In *Libraries as Wisdom Hubs: Interdisciplinary Insights for the Future*. W.S. College of Commerce, Wardha, Maharashtra.

Verma, S. (2017). Bridging theoretical and applied statistics using visual aids in competitive exam preparation. *Journal of Educational Research in India*, *12*(3), 45–55.

Wild, C. J., & Pfannkuch, M. (1999). Statistical thinking in empirical enquiry. *International Statistical Review*, *67*(3), 223–248. DOI: 10.1111/j.1751-5823.1999.tb00442.x

Chapter 11
Students' Attitudes Toward Statistics and Probabilities

Athanasia Vrettou Gavrillidi
https://orcid.org/0000-0001-6404-5003
Hellenic Open University, Greece

ABSTRACT

Courses such as probability cause anxiety in students even today. Anxiety along with other negative feelings such as fear and anxiety hinder success in learning. They additionally hinder plans for future engagement with the subject of probabilities. In the present research we study students' attitudes about probabilities. 162 students, of which 69 were female students and 93 male students completed a questionnaire assessing their attitudes towards the course. The results showed that students have very negative attitudes connected with the anxiety they feel about statistics and probability, and they consider that although they are very useful subjects and they consider that these subjects are very useful for research.

1. INTRODUCTION

The concepts of probability theory are one of the most basic subjects and branches of mathematics and statistics as it finds applications in many scientific disciplines such as business administration, economics, biology, medicine, agronomy, education, engineering, etc. Thus, the concepts of probability theory are fundamental to the understanding and analysis of complex problems (Anastasiadou, 2005; 2002; 2004a; 2004b), such as statistical inference and decision-making under conditions of uncertainty and scientific inquiry (Anastasiadou, 2005a; 2005b; 2005c; 2007a;

DOI: 10.4018/979-8-3693-9934-7.ch011

2007b). For these reasons, it occupies a prominent place in the curricula of many schools and departments of higher education (Anastasiadou, & Chadjipantelis, 2008; Anastasiadou, et al., 2013a; 2013b).

However, it is acknowledged that students not only of law or humanities faculties, but also of faculties such as mathematics and engineering and medicine faced significant difficulties and cognitive barriers in understanding and applying the concepts of probability theory and statistics (Anastasiadou, 2008a; 2008b; 2008c; 2009a; 2009b; 2010; 2019a; 2019b; 2019c), often leading to negative attitudes towards these subjects (Anastasiadou & Gagatsis, 2005a; 2005b; 2007; 2017; Gagatsis et al. 1997a; 1997b), which are subjects that raise insecurity and even anxiety in students and consequently lead to low student performance (Christodoulou at al., 2017; Nicolaou et al., 2017; Petridou et al. 2017).

Research has shown that many students experience significant cognitive difficulties in understanding and applying probability theory concepts which may be based on previous negative experiences in mathematics, in their more general beliefs about the importance and value of teaching reflective concepts, and in applying these in both their personal and professional lives (Anastasiadou et al., 2008; Anastasiadou & Pappa, 2019; Chadjipantelis & Anastasiadou, 2010). Attitudes carried over by students from previous experiences can be a hindrance to learning concepts of probability theory and reflective concepts in general (Anastasiadou & Papadimitriou, 2001a; 2001b; 2003).

The study of students' attitudes toward concepts in probability theory is very important because they are directly related to students' academic achievement and subsequent ability to apply the reflective concepts in their professional lives.

The purpose of this thesis is to explore students' attitudes towards concepts of probability theory attitudes associated with the usefulness of probability theory, anxiety, self-confidence, liking, and motivation.

2. METHODOLOGY

In the presents section, the research questions, the statistical methodology, the research instrument as well as the sample characteristics are going to be presented.

2.1 Research Questions

The study poses the following research questions:

RQ1: What are the dimensions of Likeness regarding the liking of working with statistics and probabilities?

RQ2: What are the dimensions of Anxiety regarding the fear that students demonstrate toward statistics and probabilities?

RQ3: What are the dimensions of Confidence regarding the feeling of confidence of the skill in statistics?

RQ4: What are the dimensions of Motivation regarding students feeling regarding the usefulness of statistics?

RD5: What are the dimensions of Usefulness regarding the value that students give to statistics and probabilities related to their professional career?

2.2 Statistical Methodology

Descriptive statistics were used for student responses statistical analysis. Data analysis and verification of the test's reliability were performed using SPSS software, version 25.

2.3 Participants-Sample

The following table presents the Demographic Characteristics of the Pilot Sample (Table 1).

Table 1. Demographic characteristics of the pilot sample

Demographic Characteristics	Category	Frequency N	Relative Frequency (%)
Gender	Male	93	57.4
	Female	69	42.6
Age	18-20	38	23.5
	20-22	34	21.0
	22-24	56	34.6
	>24	34	21.0
Place of Residence	Urban area	70	43.2
	Semi-urban area	65	40.1
	Rural area	27	16.7
School of Study	Engineering	17	10.5
	Natural Sciences	22	13.6
	Health Sciences	43	23.6
	Economic and Social Sciences	64	39.5
	Humanities	16	9.9

continued on following page

Table 1. Continued

Demographic Characteristics	Category	Frequency N	Relative Frequency (%)
Year of Studies	1st year	17	10.5
	2nd year	18	11.1
	3rd year	46	28.4
	4rd year	39	24.1
	5th year	23	14.2
	6th year	11	6.8
	Graduating (final year)	5	3.1
	Postgraduate or doctoral study program	3	1.9

2.4 Instrument

The questionnaire used in the study was the instrument made by Auzmendi (1992). It consists of 25 items and five conceptual constructs named "Utility/ Usefulness", "Anxiety", "Confidence/Trust", "Likeness" and Motivation". The conceptual construct named "Utility/Usefulness" consist of 5 items (Usf1: I regard Statistics and Probability as much-need matters in my career, Usf2: I want to have a deeper knowledge of Statistics and Probability, Usf3: Knowing how to use Statistics and Probability would increase my chances of working, Usf4: For the professional development of my career, I believe there are other issues that are more important than Statistics and Probability, Usf5: For the professional development of my career, one of the most important subjects to be studied in Statistics and Probability).

The conceptual construct named "Anxiety" consist of 5 items (Anx1: Statistics and Probability issues leave bad, Anx2: Statistics and Probability are topics that I'm afraid of the most, Anx3: When I come across a statistical or probabilistic problem, I cannot think clearly, Anx4: Working with Statistics and Probability makes me very nervous, Anx5: Carrying statistics makes me uncomfortable and nervous).

The conceptual construct named "Confidence/Trust" consist of 5 items (Con1: Studying or working with Statistics and Probability does not scare me, Con2: I have confidence in myself when I face statistical and probabilistic problems, Con3: I calm down when I am solving a statistical or probabilistic problem, Con4: I do not change my behavior when working with statistics and probability, Con5: If I worked, I think I would dominate the statistical and probabilistic contents).

The conceptual construct named "Likeness" consist of 5 items (Lik1: Using elements of Statistics and Probability is fun for me, Lik2: I have fun talking to others about Statistics and Probability, Lik3: The statistics and odds are nice and stimulating for me, Lik4: I would like to have a profession in which I would have to use Statistics and Probability, Lik5: If I had the opportunity, I would enroll in other courses in Statistics and Probability).

The conceptual construct named "Motivation" consist of 5 items (Mot1: Statistics and Probability are too theoretical to be used in my professional practice), (Mot2: Statistics and Probability may be useful to those involved in research but not to other professionals), (Mot3: I hope to use little statistical or probability in my professional life, Mot4: I have great satisfaction in solving statistical and probabilistic problems, Mot5: The subject that is taught in the Statistics and Probability class is very uninteresting).

2.5 Reliability Evaluation of the Research Instrument

The reliability of the instrument "Attitude toward statistics (SATS) of Auzmendi (1992)" was evaluate by Cronbach alpha coefficient (*a*). Cronbach alpha coefficient (*a*) values above 0.80 are considered to be satisfactory (Anastasiadou & Dimitriadou 2011; Anastasiadou & Tiliakou, 2016; Anastasiadou, & Taraza, 2019; Anastasiadou, 2019; Anastasiadou & Karakos, 2011).

The reliability coefficient Cronbach *a* of conceptual constructs named "Utility/Usefulness", "Anxiety" and "Confidence/Trust" count for 0.765, 0.705 and 0.784 and they considered to be quite satisfactory (Anastasiadou et al., 2010; Papadaki & Anastasiadou, 2019; Fotiadis & Anastasiadou, 2018a; 2018b; Taraza & Anastasiadou, 2019; Anastasiadou, 2013; 2016; 2019d; Anastasiadou et al., 2014; 2022; 2023; Papadaki et al., 2023; Anastasiadou, 2023).

The reliability coefficient Cronbach *a* of conceptual constructs named "Likeness" and "Motivation" values for 0.841 and 0.871 and they are considered to be satisfactory (Anastasiadou & Masouras, 2023; Anastasiadou et al., 2023; Anastasiadou & Zrinoglou, 2025), Anastasiadou et al., 2004; Seremeti et al., 2024; Tataza et al., 2024; Valkanos et al. 2010; Paniistides et al., 2015; Anastasiadou at al. 2022; Valsamidis et al., 2024; Kapetanopoulu et al., 2021; Anastasiadou et al., 2022; Anastasiadou et al, 2022) (Table 2).

Table 2. Reliability evaluation

Reliability evaluation	**Cronbach *a***
Instrument "Attitude toward statistics (SATS) (Auzmendi, 1992)	0.907
Utility/Usefulness	0.765
Anxiety	0.705
Confidence/Trust	0.784

continued on following page

Table 2. Continued

Reliability evaluation	**Cronbach *a***
Likeness	0.841
Motivation	0.871

RESULTS

The following table (Table 3) presents the percentage of students' responses.

Table 3. Items percentage

Conceptual Construct	Code	Items	Total disagree	Disagree	Neither disagree nor agree	Agree	Total agree
Utility/ Usefulness	Usf1	I regard Statistics and Probability as much-need matters in my career.	0.6%	2.5%	4.9%	40.1%	51.9%
	Usf2	I want to have a deeper knowledge of Statistics and Probability	0.6%	2.5%	4.3%	37%	55.6%
	Usf3	Knowing how to use Statistics and Probability would increase my chances of working	0%	3.1%	8%	29%	59.9%
	Usf4	For the professional development of my career, I believe there are other issues that are more important than Statistics and Probability.	0%	1.9%	17.9%	42%	38.3%
	Usf5	For the professional development of my career, one of the most important subjects to be studied in Statistics and Probability	0.6%	8.6%	24.1%	33.3%	33.3%

continued on following page

Table 3. Continued

Conceptual Construct	Code	Items	Total disagree	Disagree	Neither disagree nor agree	Agree	Total agree
Anxiety	Anx1	Statistics and Probability issues leave bad	1.9%	2.5%	10.5%	37%	48.1%
	Anx2	Statistics and Probability are topics that I'm afraid of the most	0.6%	5.6%	12.3%	32.7%	48.8%
	Anx3	When I come across a statistical or probabilistic problem, I cannot think clearly.	10.5%	16%	36.4%	24.1%	13%
	Anx4	Working with Statistics and Probability makes me very nervous	29.6%	19.8%	21%	16%	13.6%
	Anx5	Carrying statistics makes me uncomfortable and nervous	17.8%	19.2%	32.2%	20.2%	10.6%
Confidence/ Trust	Con1	Studying or working with Statistics and Probability does not scare me.	18.5%	18.5%	36.4%	19.8%	6.8%
	Con2	I have confidence in myself when I face statistical and probabilistic problems	4.3%	12.3%	30.2%	36.4%	16.7%
	Con3	I calm down when I am solving a statistical or probabilistic problem	8.6%	11.7%	43.8%	26.5%	9.3%
	Con4	I do not change my behavior when working with statistics and probability	8%	14.8%	30.9%	34.6%	11.7%
	Con5	If I worked, I think I would dominate the statistical and probabilistic contents	4.9%	11.1%	22.8%	42.6%	18.5%
Likeness	Lik1	Using elements of Statistics and Probability is fun for me	5.6%	16.7%	37%	32.7%	8%
	Lik2	I have fun talking to others about Statistics and Probability	2.5%	10.5%	13.6%	32.7%	40.7%
	Lik3	The statistics and odds are nice and stimulating for me.	1.2%	4.9%	16%	42.6%	35.2%

continued on following page

Table 3. Continued

Conceptual Construct	Code	Items	Total disagree	Disagree	Neither disagree nor agree	Agree	Total agree
	Lik4	I would like to have a profession in which I would have to use Statistics and Probability	4.3%	14.8%	23.5%	41.4%	16%
	Lik5	If I had the opportunity, I would enroll in other courses in Statistics and Probability	1.2%	9.9%	30.9%	38.9%	19.1%
Motivation	Mot1	Statistics and Probability are too theoretical to be used in my professional practice	2.5%	13%	29.6%	37.7%	17.3%
	Mot2	Statistics and Probability may be useful to those involved in research but not to other professionals.	3.7%	13%	25.3%	37.7%	20.4%
	Mot3	I hope to use little statistical or probability in my professional life.	4.9%	8%	20.4%	35.2%	31.5%
	Mot4	I have great satisfaction in solving statistical and probabilistic problems	5.6%	14.8%	37%	32.1%	10.5%
	Mot5	The subject that is taught in the Statistics and Probability class is very uninteresting	5.6%	7.4%	25.3%	44.4%	17.3%

From the results of the students' answers it was made clear that the 92% of them claimed that Statistics and Probability as much-need matters in their career, the 92.6% wanted to have a deeper knowledge of Statistics and 88.9% wanted to Know how to use Statistics and Probability would increase their chances of working even though 80.3% believed that for the professional development of their career, I believe there are other issues that are more important than Statistics and Probability. 66.6% believed that for the professional development of their career, one of the most important subjects to be studied in Statistics and Probability.

As far as "Anxiety" conceptual construct concerns, 85.1% think that Statistics and Probability issues leave bad and 81.5% Statistics and Probability are topics that they are afraid of the most.

As far as “Confidence” conceptual construct concerns, 53.1% asserted that they have confidence in themself when they face statistical and probabilistic problems. In addition, 61.1% asserted that if they worked, they think they would dominate the statistical and probabilistic contents.

Regarding “Likeness” conceptual construct, 73.4% claimed that they have fun talking to others about Statistics and Probability, 77.8% considered that statistics and odds are nice and stimulating for me, 56.4% they would like to have a profession in which they would have to use Statistics and Probability, while 58% admitted that if they had the opportunity, they would enroll in other courses in Statistics and Probability.

Regarding “Motivation'' conceptual construct, 58.1% claimed that Statistics and Probability may be useful to those involved in research but not to other professionals, 66.7% hoped to use little statistical or probability in my professional life, 61.7% claimed the subject that is taught in the Statistics and Probability class is very uninteresting.

CONCLUSIONS

Students’ attitudes toward statistics and probabilities are subject of major importance in statistical education. Teaching methodologies must pay attention to those attitudes, maybe a very be a serious obstacle in learning statistics and probability. Understanding students' behaviors and attitudes toward statistics and probabilities requires further research, particularly leveraging methods involving stochastics concepts (Souravlas & Anastasiadou, 2022; Souravlas et al., 2020), statistical inference and decision-making under conditions of uncertainty and scientific inquiry (Anastasiadou, 2005a; 2005b; 2005c; 2007a; 2007b), big data, Markov models, cloud computing technologies (Souravlas et al., 2021a; 2021b; 2021c), advanced technological solutions, and novel educational methods. Conducting this research is essential due to the importance of thoroughly exploring the emotional and cognitive dimensions related to stochastics concepts (Souravlas et al., 2022a; 2022b; 2023).

REFERENCES

Anasatsiadou, S., & Papadimitriou, I. (2003). Secondary mathematics teachers' attitudes towards statistics in Greece. 3° Mediterranean Conference in Mathematics Education, Athens, pp. 441-448.

Anastasiadou, S. (2002). Enhancing prospective teachers' knowledge of students' attitudes and representations towards statistics-the case of Greek students. In the Journal *'Scientia Paedacogika Experimentals'. XXXIX, 1*, 103–127.

Anastasiadou, S. (2004a). Teacher's opinions about statistics and probability. In the journal *'Quaderni di Ricerca in Didattica'*, n_o 14. pp. 134-143. http://math.unipa .it/~grim/quaderno14.htm

Anastasiadou, S. (2004b). Perceptions-Attitudes-Conducts of the Greek Mathematicians for Statistics in Secondary Education, D. De Bock, M. Isoda, J. A. C. Cruz, A. Gagatsis, E Simmt (eds) 10th International Congress on Mathematical Education (ICME 10) Copenhagen, Demark, pp. 61-68.

Anastasiadou, S. (2005). Analysis of compound variables concerning students' attitudes towards statistics after the first contact with the lesson of statistics. In the journal *'Meditereeanean. Journal for Research in Mathematics Education, 4*(1), 71–102.

Anastasiadou, S. (2005a). Survey of attitudes and beliefs toward Statistics in Greek Secondary Education International Conference on Mathematics Education, Svishtov, Bulgaria, pp. 125-133.

Anastasiadou, S. (2005b). Affective reactions and attitudes of the last class of Greek high school students towards statistics, In M Boosch (Eds) Proceedings 0r the Fourth Conference of the European Society for Research in Mathematics Education (CERME 4), Sant Feliu de Guíxols, Spain, 2005 in the http://cerme4.crm.es/Papers %20definitius/5/Anastasiadou%20Sofia.pdf

Anastasiadou, S. (2005c). Affective reactions and attitudes of the last class of Greek high school students towards statistics. Proceedings of CERME IV, European Research in Mathematics Education. Sant Feliu de Guíxols, Girona: CERME.

Anastasiadou, S. (2007a). The Role of representations in solving statistical problems and the translation ability of fifth and sixth grade students. *International Journal of Learning, 14*(10), 125–132. DOI: 10.18848/1447-9494/CGP/v14i10/45497

Anastasiadou, S. (2007b). Affective Responses and Attitudes of ninth grade Greek Students. *Proceedings of the 5th Mediterranean Conference in Mathematics Education*, Rodes, Greece pp. 169-178.

Anastasiadou, S. (2008a). Iconic Representations and Statistical Problem Solving: The Case of Greek Pupils. *International Journal of Learning, 15*(11), 301–310. DOI: 10.18848/1447-9494/CGP/v15i11/58699

Anastasiadou, S. (2008b). Representations and learning in statistics: a comparative study between Greek primary school students and immigrants. Eleventh International Congress on Mathematical Education, in its eleventh edition is to be held in Monterrey, Mexico, 2008 in the http://tsg.icme11.org/document/get/195

Anastasiadou, S. (2008c). The Effects of representational systems on the learning of statistics between Greek primary school students and immigrants. Proceeding of the 5th International Mathematics Education and Society Conference (MES 5) Albufeira, Portugal, pp.167-176.

Anastasiadou, S. (2009a). A structural equation model representing the hierarchical structure of translations among representations in Frequency concept. In Novotha, J., & Moraova, H. (Eds.), *International Symposium Elementary Mathematics Teaching (SEMT 09)* (pp. 57–64). Czech Repuplic., https://storage.cet.ac.il/cetnews/SEMT09.PDF

Anastasiadou, S. (2009b). Greek students' ability in probability problem solving. In the Sixth Conference of the European Society for Research in Mathematics Education. Proceedings of CERME 6, Lyon France © INRP 2010 <www.inrp.fr/editions/cerme6> pp. 404-412. https://ife.ens-lyon.fr/editions/editions-electroniques/cerme6/working-group-3

Anastasiadou, S. (2010). Pre-service teachers' performance on the learning of probability distributions and the role of projects: A multilevel statistical analysis. In Regnier, Jean-Claude, Spagnolo, Filippo, Di Paola, Benedetto, & Gras, Regis, Eds, *5th International Conference Implicative Statistic Analysis*, Palermo, Italy *pp* 73-84.

Anastasiadou, S. (2013). Developing and evaluating a structural equation model measuring leadership changes in a lifelong learning world. *The International Journal of Educational Organization and Leadership, 19*(2), 1–17. DOI: 10.18848/2329-1656/CGP/v19i02/48520

Anastasiadou, S. (2016). *Evaluation of the implementation of TQM principles in tertiary education using the EFQM excellence model: Research in educational departments of Greek universities* [Master's thesis, Greek Open University]. Patra, Greece.

Anastasiadou, S. (2019a). *Representations in statistics problem solving.ICERI2019, the* 12th annual International Conference of Education, Research and Innovation, Seville (Spain), ICERI2019, pp. 8993- 8996. DOI: 10.21125/iceri.2019.2159

Anastasiadou, S. (2019b). Internal representations toward external representations: The case of statistical concepts. 11th annual International Conference on Education and New Learning Technologies Palma de Mallorca, Spain, EDULEARN 19, pp. 6-9.

Anastasiadou, S. (2019c). Pre-service teachers' attitudes toward the use of different modes of representation in statistics teaching and learning. 11th annual International Conference on Education and New Learning Technologies Palma de Mallorca, Spain, EDULEARN 19. pp. 1-5.

Anastasiadou, S. (2019d). Comparison of contemporary advanced statistical methods regarding construct validity evaluation of TEIque-SF instrument: Statistical implicative analysis vs. principal components analysis. In *9ème Colloque International sur Analyse Statistique Implicative (ASI 10)* (pp. 148–163). Belfort, France.

Anastasiadou, S. (2023). Psychometric Testing Procedures for Attitudes Toward Reproductive Tourism and Cross Border Reproductive Care (ATCBRC). In Kavoura, A., Borges-Tiago, T., & Tiago, F. (Eds.), *Strategic Innovative Marketing and Tourism. ICSIMAT 2023. Springer Proceedings in Business and Economics*. Springer., DOI: 10.1007/978-3-031-51038-0_14

Anastasiadou, S., Anastasiadis, L., Angeletos, T., & Vandikas, J. (2010). A Multidimensional Statistical Analysis of Students' Attitudes toward Physics. *The International Journal of Diversity in Organisations, Communities and Nations, 16*(5), 341–356. DOI: 10.18848/1447-9532/CGP/v10i01/39794

Anastasiadou, S., Chadjipadelis, T., & Kofou, I. (2013a). Pre-school Education department to understand probability distributions. International Symposium Elementary Mathematics Teaching (SEMT '13), Prague, Czech Republic, In Jarmila Novotná and Hana Moraová ed., pp 67-75.

Anastasiadou, S., Chadjipadelis, T., & Kofou, I. (2013b). The role of projects in helping Greek university students from Pre-school Education department to understand probability distributions International Symposium Elementary Mathematics Teaching (SEMT '13). Prague, Czech Repuplic. International Symposium Elementary Maths Teaching. Proceedings: Tasks and tools in elementary mathematics, Prague [2013] σελ.67-74

Anastasiadou, S., & Chadjipantelis, T. (2008). The role of representations in the understanding of probabilities in tertiary education. Eleventh International Congress on Mathematical Education, Monterrey, Mexico, 2008 in the http://tsg.icme11.org/tsg/show/14

Anastasiadou, S., Elia, I., & Monoyiou, A. (2008). The effects of representations on the learning and understanding of statistics in Greek primary school. *Mediterranean Journal for Research in Mathematics Education*, *7*(1), 67–80.

Anastasiadou, S., & Gagatsis, A. (2005a). Attitudes vers la statistique des étudiants –futurs enseignants de l'école primaire grecque. In Gras, Regis, Spagnolo, Filippo, & David, Jerome Eds, *3nd International Conference Implicative Statistic Analysis*, Palermo, Italy, pp. 109-118.

Anastasiadou, S., & Gagatsis, A. (2005b). Believes and academic behaviors in statistics education. In Novotha, J. Eds, *International Symposium Elementary Mathematics Teaching SEMT 05 Prague*, pp. 59-65.

Anastasiadou, S., & Gagatsis, A. (2007). Exploring the effects of representation on the learning of statistics in Greek primary school. In the Fifth Conference of the European Society for Research in Mathematics Education (CERME 5), Nikosia, Cyprus pp. 90-100. http://ermeweb.free.fr/CERME5b/ και http://ermeweb.free.fr/CERME%205/WG1/1_Anastasiadou.pdf

Anastasiadou, S., & Gagatsis, A. (2017). Students' representations of linear motion. *Proceedings* of 9ème Colloque International sur Analyse Statistique Implicative. 9th International Meeting Statistical Implicative Analysis. Belfort – France, pp. 479-494.

Anastasiadou, S., & Masouras, A. (2023). Exploring determinants regarding Visit and Revisit intentions considering Greece as a Medical Tourism destination. 10th International Conference on Strategic Innovative Marketing and Tourism. Zakynthos, Ionian Islands, Greece. In A. Kavoura et al. (eds.), Strategic Innovative Marketing and Tourism, Springer Proceedings in Business and Economics, (2023), 101-109. DOI: 10.1007/978-3-031-51038-0_12

Anastasiadou, S., Masouras, A., & Papademetriou, C. (2023). Attitudes toward Repro-ductive Tourism and Cross border reproductive care (CBRC): Legal, Economic, Eth-ical issues and dilemmas, possibilities and limitations. *Proceedings of the 6th Inter-national Conference on Tourism Research, ICTR 2023*, Pafos, Cyprus, pp. 17-23.

Anastasiadou, S., Masouras, A., & Papademetriou, C. (2023). Attitudes toward Reproductive Tourism and Cross border reproductive care (CBRC): Legal, Economic, Ethical issues and dilemmas, possibilities and limitations. *Proceedings of the 6th International Conference on Tourism Research, ICTR 2023*, 7-23, Pafos, Cyprus. DOI: 10.34190/ictr.6.1.1189

Anastasiadou, S., & Papadimitriou, I. (2001a). *Another way to estimate the validity of a questionnaire for measurement of attitudes towards statistics. The case of Greek students. In the Journal 'Learning in Mathematics and Science and Educational Technology* (Vol. II). Intercollege Press.

Anastasiadou, S., & Papadimitriou, I. (2001b). *Another way to estimate the validity of a questionnaire for measurement of attitudes towards statistics. The case of Greek students. In the Journal 'Learning in Mathematics and Science and Educational Technology* (Vol. II). Intercollege Press.

Anastasiadou, S., Papalexandris, S. & Konteos, G. (2022). The Effect of Emotional Intelligence on Perceived e-Service Quality, Consumers' Perceived Value, Purchase, Loyalty Intentions and Satisfaction, Haicta. CEUR Workshop Proceedings (CEUR-WS.org), 151-161.

Anastasiadou, S., & Pappa, A. (2019). *Greek pre-service teachers' perceptions, emotions and attitudes toward representations of physic concepts.* ICERI2019, the 12th annual International Conference of Education, Research and Innovation, Seville (Spain), ICERI2019, pp. 8987-8992. DOI: 10.21125/iceri.2019.2158

Anastasiadou, S. Zirinoglou, P. (2025). L'Analyse Factorielle des Correspondances Reveal Students' Attitudes Toward Entrepreneurship Source Title: Real-World Tools and Scenarios for Entrepreneurship Exploration, (2025) 79-100. .DOI: 10.4018/979-8-3693-3100-2.ch004

Anastasiadou, S. Zirinoglou, P. Seremeti, L. Masouras, A. Komodromos, M. Anastasiadis, L. (2024). Statistical Implicative Analysis: A Distinguished Statistical Methodology in Evaluating Greek Students' Attitudes Toward the Use of AI Source Title: Real-World Tools and Scenarios for Entrepreneurship Exploration, (2024), 79-100, .DOI: 10.4018/979-8-3693-3100-2.ch006

Anastasiadou, SD. Anastasiadis, L. Florou, G. & Valsamidis, S. (2022). Greek students' Environmental Attitude based on the Theory of Ecological Attitude. HAICTA, (2022), 222-231.

Anastasiadou, S. D., Batiou, V., & Balkanos, E. (2014). Occupation mobility in Greece: A completely new area. Global Journal of Business. *Economics and Management*, *4*(2), 73–80.

Auzmendi, E. (1992). *Las actitudes hacia la matemática-estadística en las enseñanzas medias y universitarios*. Mensajero.

Chadjipantelis, T., & Anastasiadou, S. (2010). Pre-service teachers' understanding of probability Distributions: A Multilevel Statistical Analysis. International Conference on the Teaching of Statistics. ICOTS 8, Ljubljanja, Slovenia 2010. https://www.stat.auckland.ac.nz/~iase/publications.php?show=icots8

Christodoulou, T., Gagatsis, A., & Anastasiadou, S. (2017). Enquête internationale comparative: croyances et attitudes des élèves du secondaire inférieur relatives aux erreurs en mathématiques. *Proceedings* of 9ème Colloque International sur Analyse Statistique Implicative, Belfort – France, pp. 404-418.

Fotiadis, Th., & Anastasiadou, S. (2018a). Contemporary advanced statistical methods for the science of marketing: Implicative Statistical Analysis vs Principal Components Analysis. Proceeding of the ICME 2018- 1st International Conference on Marketing and Entrepreneuship. Cyprus. https://hephaestus.nup.ac.cy/handle/11728/11393

Fotiadis, Th., & Anastasiadou, S. (2018b). Contemporary advanced statistical methods for the science of marketing: Principal Components Analysis vs Analyse Factorielle des Correspondances. In *Proceedings of the ICME 2018 – 1st International Conference on Marketing and Entrepreneurship*. Cyprus. Retrieved from https://hephaestus.nup.ac.cy/handle/11728/11392

Gagatsis, A., Anastasiadou, S., & Mpora, E. (1997a). Errors made by Greek students of mathematics in probabilities, in the Journal '*Didactics of mathematics-Technology in Education',* pp. *47-66.*

Gagatsis, A., Anastasiadou, S., & Mpora, E. (1997b). Errori commess; da studenti greci di Matematica in questioni di probabilita, In the journal '. *Bollettino dei Docenti di Matematica, 35*, 87–102.

Kapetanopoulou, P., Kouroutzi, A., & Anastasiadou, S. (2021). The Impact of Information Systems Implementation in the Greek Manufacturing Enterprises. *Applied Sciences (Basel, Switzerland), 11*(24), 11781. DOI: 10.3390/app112411781

Nicolaou, S., Gagatsis, A., Deliyianni, E., Panaoura, A., Elia, I., & Anastasiadou, S. (2017). Tracing the beliefs and self-efficacy beliefs of undergraduate economic sciences students: the case of the representations of functions. *Proceedings* of 9ème Colloque International sur Analyse Statistique Implicative. Belfort – France, pp. 461-478.

Panitsides, E., & Anastasiadou, S. (2015). Lifelong Learning Policy Agenda in the European Union: A bi- level analysis. *Open Review of Educational Research*, *2*(1), 128–142. DOI: 10.1080/23265507.2015.1043936

Papadaki, Z., Anastasiadou, S., Masouras, A., & Papalexandris, S. (2023). Factors Influencing the Use of Digital Marketing by Telemedicine Services. 10th International Conference on Strategic Innovative Marketing and Tourism. Zakynthos, Ionian Islands, Greece. In A. Kavoura et al. (eds.), Strategic Innovative Marketing and Tourism, Springer Proceedings in Business and Economics, (2023), 855-862. DOI: 10.1007/978-3-031-51038-0_92

Papadaki, Z. E., & Anastasiadou, S. D. (2019). Evaluating perception, expectation of consumers, and service quality gap in Greek banking in a period of financial crisis and capital controls. In N. Sykianakis, P. Polychronidou, & A. Karasavvoglou (Eds.), Economic and Financial Challenges for Eastern Europe (pp. 67–80). Springer Pro-ceedings in Business and Economics. Springer, Cham. https://doi.org/DOI: 10.1007/978-3-030-12169-3_5

Petridou, A., Elia, I., Gagatsis, A., & Anastasiadou, S. (2017). L'appréhension perceptive des figures géométrique. *Proceedings* of 9ème Colloque International sur Analyse Statistique Implicative. Belfort – France, pp. 231-247.

Seremeti, L., Anastasiadis, L., & Anastasiadou, S. (2024). The role of data in Environmental, Social and Governance strategy. In *Papademetriou, C., Anastasiadou, S. Ragazou, K., Garefalakis, A., and Papalexandris, S. (2024)* (pp. 201–218). ESG and Total Quality Management in Human Resources. IGI Global Publisher of Timely Knowledge., DOI: 10.4018/979-8-3693-2991-7.ch010

Souravlas, S., & Anastasiadou, S. (2020). Pipelined dynamic scheduling of big data streams. *Applied Sciences (Basel, Switzerland)*, *10*(14), 4796. DOI: 10.3390/app10144796

Souravlas, S., & Anastasiadou, S. (2022). On implementing social community clouds based on Markov models. *IEEE Transactions on Computational Social Systems, 2022*. Advance online publication. DOI: 10.1109/TCSS.2022.3213273

Souravlas, S., Anastasiadou, S., Economides, Th., & Katsavounis, S. (2023). Probabilistic community detection in social networks. *IEEE Access : Practical Innovations, Open Solutions*, *11*, 25629–25641. DOI: 10.1109/ACCESS.2023.3257021

Souravlas, S., Anastasiadou, S., & Katsavounis, S. (2021b). A survey on the recent advances of deep community detection. *Applied Sciences (Basel, Switzerland)*, *11*(16), 7179. DOI: 10.3390/app11167179

Souravlas, S., Anastasiadou, S., & Katsavounis, S. (2021c). More on pipelined dynamic scheduling of big data streams. *Applied Sciences (Basel, Switzerland), 11*(1), 61. DOI: 10.3390/app11010061

Souravlas, S., Anastasiadou, S., & Kostoglou, I. (2022b). A novel method for general hierarchical system modeling via colored Petri nets based on transition extractions from real data sets. *Applied Sciences (Basel, Switzerland), 13*(1), 339. DOI: 10.3390/app13010339

Souravlas, S., Anastasiadou, S., Tantalaki, N., & Katsavounis, S. (2021a). A dynamic load balanced task distribution strategy for heterogeneous cloud platforms based on Markov process modeling. *IEEE Access : Practical Innovations, Open Solutions, 10*, 2169–3536. https://ieeexplore.ieee.org/abstract/document/9729723

Souravlas, S., Anastasiadou, S. D., Tantalaki, N., & Katsavounis, S. (2022a). A fair, dynamic load balanced task distribution strategy for heterogeneous cloud platforms based on Markov process modeling. *IEEE Access : Practical Innovations, Open Solutions, 10*, 26149–26162. DOI: 10.1109/ACCESS.2022.3157435

Souravlas, S., Katsavounis, S., & Anastasiadou, S. (2020). On modeling and simulation of resource allocation policies in cloud computing using colored Petri nets. *Applied Sciences (Basel, Switzerland), 10*(16), 5644. DOI: 10.3390/app10165644

Taraza, E., & Anastasiadou, S. (2019). EFQM Excellence Model in Vocational Lyceum: Reliability and validity of the EFQM instrument. In *Proceedings of the 13th Annual International Technology, Education and Development Conference (INTED2019)(pp.* 2273–2285*)*. Valencia, Spain. DOI: 10.21125/inted.2019.0637

Taraza, E., Anastasiadou, S., Seremeti, L., & Masouras, A. (2024). Environmental, Social, and Governance of Businesses and Organizations and the Role of Human Resources. In *Papademetriou, C., Anastasiadou, S. Ragazou, K., Garefalakis, A., and Papalexandris, S. (2024)* (pp. 116–132). ESG and Total Quality Management in Human Resources. IGI Global Publisher of Timely Knowledge., DOI: 10.4018/979-8-3693-2991-7.ch006

Taraza, E., Papademetriou, C., Seremeti, L., Masouras, A., & Anastasiadou, S. (2024). The Necessity of Environmental Ethics in Environmental Education and the Role of Educators. In *Papademetriou, C., Anastasiadou, S. Ragazou, K., Garefalakis, A., and Papalexandris, S. (2024)* (pp. 149–166). ESG and Total Quality Management in Human Resources. IGI Global Publisher of Timely Knowledge., DOI: 10.4018/979-8-3693-2991-7.ch008

Valkanos, E., Giossi, S., & Anastasiadou, S. (2010). Certified Adult Educators´ Attitudes towards Continuing Education. Review of European Studies. 2(1), (2010) 61-68. Part of ISSN: 19187173 19187181.

Valsamidis, S., Mandilas, A., Florou, G., & Anastasiadou, S. (2024). The use of e-government services provided by the local authorities: The case of the municipality of Kavala. *International Journal of Electronic Governance, 16*(2), 196–218. DOI: 10.1504/IJEG.2024.140790

Chapter 12
The Role of Different Modes of Representation in Statistics

Athanasia Vrettou Gavrillidi
https://orcid.org/0000-0001-6404-5003
Hellenic Open University, Greece

ABSTRACT

This paper investigates the effectiveness of teaching different kinds of representations in teaching probabilities. Verbal, algebraic and graphical representations are used in teaching Venn diagrams. A sample of 127 students was used to examine their effectiveness in teaching and learning probability concepts. Implicative statistical analysis is applied to evaluate the relation between students' approach and their ability to solve problems. More exactly, this study investigates 127 students' abilities in using representations (verbal, tabular, graphical and symbolic form) of basic statistical concepts and in moving from one mode of representation to another successfully. Diagrammatic modes seem to be easier than the other nodes and are linked to higher performance.

INTRODUCTION

The aim of this paper is to study the absolute efficiency of the use of different types of representations relating to the understanding of probability concepts. In particular, the aim is to explore the use of 3 different representation approaches and more specifically the vocabulary of algebraic and graphic representation as well as table representation in teaching Venn diagrams and concepts related to probability theory. According to the international literature the use of different types of repre-

DOI: 10.4018/979-8-3693-9934-7.ch012

sentations in the teaching of mathematics in general and statistics and probability in particular is very important for the understanding of the concepts about it and this paper investigates the effectiveness of the use of these different types of representations and the effortlessness with which they are used (Anastasiadou, 2007a; 2009a; 2009b).

The value of representations has been emphasized by many eminent researchers (Kaput 1987; Ainsworth et al., 1997; Duval, 2002; Dufur-Janvier et al., 1987). In addition, the use of the term 'mathematics' in the context of the teaching of mathematics is not restricted to the teaching of mathematics (DeWindt-King & Goldin, 2003; Goldin 1998). In addition to the teaching of statistics and concepts of probability theory. As researchers typically point out, an understanding of a concept such as a mathematical or statistical concept or a concept of probability theory becomes apparent when one successfully transitions from one form of representation to another (Ainsworth et al. 1997; Ainsworth, 2006; Janvier, 1987; Duval, 1987; 2002; Chadjipantelis & Anastasiadou, 2010).

Venn diagrams are used a lot in understanding probability theory because they help in understanding set theory and create positive attitudes towards probability theory and statistics (Anastasiadou, 2002; 2005; 2008a; 2008b; Anastasiadou, 2008c). Venn diagrams were introduced in 1880 by John Venn (1834-1923), intended to show the different ways in which concepts are represented among a collection of sets and to illustrate their relationship in probability theory, mathematical logic, statistics, linguistics, and computer science.

The importance of the use of Venn diagram representations is extremely important (Anastasiadou, 2004a; Anastasiadou & Chadjipantelis, 2008, Anastasiadou et al, 2013a; 2013b). These diagrams are essentially a very important cognitive tool and more specifically a very powerful visual cognitive tool that allows for the visualization of simple and complex films (Anastasiadou, 2004b; Anastasiadou & Gagatsis, 2005a; 2005b). More specifically it allows the visualization of abstract concepts and can with great effortless it contributes to the simplification of complex relationships that relate to set relationships used in the definition and calculation of probability theory (Anastasiadou & Gagatsis, 2007; 2017).

In addition, the importance of using Venn diagrams focuses on simplifying complex concepts by providing the possibility for complex relationships between sets to be simplified with the help of a clearer and more intuitive visual representation of simple sets and their relationships. The union, the intersection, and the complementarity involved in set theory are much better understood with the help of graphical representations (Chadjipantelis & Anastasiadou, 2010).

Visualization helps in understanding them in depth thus enhancing the cognitive process as it is much easier for students to understand with greater clarity the concepts described symbolically and consequently and control thus enhancing the ability to solve more complex probability problems and statistical problems (Anastasiadou, 2005a; 2005b; 2005c). Graphic representation and visualization work positively in understanding even the most difficult symbolic representations, thus maximizing the understanding of students who develop more positive attitudes for stochastic concepts (Anastasiadou, 2007b; 2010; Christodoulou et al., 2017; Anastasiadou & Pappa, 2019) and consequently increasing their success in resolving the corresponding problems (Anastasiadou, 2019a; 2019b). However, each student is allowed to approach problems with the type of representation that facilitates him or her, in his own way that he or she learns and greater ease to reach the highest possible performance (Anastasiadou, S. 2019c; Anastasiadou et al., 2008; Gagatsis et al., 1997a). In conclusion, the use of Venn charts in statistical education and more generally to learning the theory of probability contributes decisively to greater and more meaningful understanding of complex problems by exaggerating the difficulties and cognitive obstacles it encounters (Gagatsis et al., 1997b; Nicolaou, et al., 2017; Deliyanni et al., 2017) developing positive attitudes towards the cognitive objects of reflective concepts (Anastasiadou & Papadimitriou, 2001a; 2001b; 2003).

The main cognitive obstacles associated with Venn diagrams refer to both the symbolic representations that relate to sets and their corresponding verbal representations. The difficulty of understanding abstract concepts is very great and more specifically relates to the abstract concepts of the intersection of two sets A and B, $A \cap B$) or the union of two sets A orB ($A \cup B$) or the complementary set A of a given set A. The more complex Venn diagrams that have depicted more difficult relationships between 2 or more sets result in cognitive difficulties on the part of the students. Another very important cognitive barrier is that which refers to the ease with which we can go from one form of representation to another, for example from the verbal to the diagrammatic or from the verbal to the symbolic or vice versa, and we always refer to Venn diagrams.

The theory of sets also seems to trouble students, and the concept of percentages, probabilities, or even difficult arithmetic operations are cognitive barriers to understanding the above concepts. The possibility of switching from one form of representation to another, especially from verbal to algebraic or symbolic and vice versa, is difficult (Anastasiadou & Chadjipantelis, 2008).

Symbolic representations $A \cup B$ or $A \cap B$ or $(A \cap B)'$ or $(A \cup B)'$ or $A \cap B'$ $orA' \cap B$ etc., are also not easily linked to the corresponding graphical representations. Still, one of the very great difficulties is that the notion of the section $A \cap B$ is connected with the two sets A and B being true at the same time, and the notion of «and» represents the section ($\cap$) and the notion of «or» represents the union ($\cup$). Students seem to have difficulty separating the concept of the section from the concept of the union both algebraically, lexically and diagrammatically. Consequently, the probability calculus of the possible represented by the totals does not seem to be easy for students (Anastasiadou & Chadjipantelis, 2008).

Therefore, the purpose of this paper is to see the effectiveness of verbal, algebraic or symbolic, diagrammatic or graphic representations and table representations in teaching and learning basic concepts of probability theory. More specifically, the purpose of this paper is to study the different representations in the successful solution of a statistical problem, or a problem related to probability theory. What is important is to explore which of the modes of representation is more understandable, and how it can contribute to the understanding of concepts.

SAMPLE OF THE RESEARCH-DOCUMENT-STATISTICAL METHODOLOGY

A sample of 127 students was selected to investigate the effectiveness of Venn diagram representations in teaching and learning probability concepts. Descriptive statistics were used to analyse the results. There was an essay 12 exercises concerning the definition of probability, Venn diagrams, and solving probability problems. The students' answers to the exercises were coded with a capital letter followed by a number of the exercise and 2 lowercase letters relating to the sense in which the exercise was carried out. The letter V stands for the Venn diagram, the P diagram for the probability problem, the letter R for the definition of the concept (e.g. event). The lower-case letters that were in the formula of representation for example r=representation, t=tabular, v=verbal, s=symbol, a=algebraic, g=graphical. The first small letter was for the initial representation and the second for the final representation. The correct answers were coded with a 1, the wrong ones with a 0, and the ones that were a few shits with a 0.5. For example, the first and second tasks are as follows:

> Task 1. Given two events A *and* B of a change experiment. Let the event $A' \cup B'$. Present this evet by a Venn diagram (encoded as $V1sg$).
>
> Task 1. Given two events A *and* B of a change experiment. Let the event $(A \cap B)' \cup (A \cup B)'$. Express this event verbally (encoded as *V2s*v).

RESULTS

The descriptive results are shown in the following Table 1.

Table 1. Students' success rate in all tasks regarding all types of conversions

Tasks	Type of translation	Success rate (%)	Tasks	Type of translation	Success rate (%)
V1sg	Symbolic-Graphic	54.2	P7va	Verbal-Algebraic	40.5
V2sv	Symbolic-Verbal	53.9	P8vg	Verbal-Graphic	36.8
V3gs	Graphic-Symbolic	37.7	P9vs	Verbal--Symbolic	32.4
V4gv	Graphic-Verbal	35.9	P10vv	Verbal-Verbal	27.6
V5vg	Verbal-Graphic	48.5	R11vv	Verbal-Verbal	27.4
V6vs	Verbal-Symbolic	52.8	R12vs	Verbal--Symbolic	21.6

From the analysis of the results, we observe that the highest success rates were those involving the translation from symbolic to graphic *V1sg* (54.2%), translation from symbolic to verbal representation *V2sv* (53.9), translation from verbal to symbolic representation of *V6vs* (52.8%) in Venn diagrams. The lowest success rates were Task Regarding the definition of concept (R). More specifically, the lowest success rate involved the translation from the verbal to the symbolic representation of R12vs (21.6%) and from the verbal to the verbal translation R11vv (27.4%).

CONCLUSION

For the study investigating the understanding of representations of concepts in probability theory, we selected a representative sample of 127 students to evaluate the effectiveness of representations in successfully solving a probability problem. The way in which students can move freely from one form of representation to the other has been examined, which also demonstrates the understanding of the concept. It was examined whether students can successfully move from vocabulary to symbolic, to graphical representation and vice versa, from vocabulary to vocabulary with regard to Venn diagrams, definitions of probability concepts as well as problems involving probability relationships. The use of diagrams did not lead to higher performance of students in the essays and it was concluded that when the initial representation is symbolic it is easier for students to correctly respond to its correspondence, in its transition more specifically to graphic and verbal.

Also, the transition from an original verbal representation to a symbolic one was equally easy. In contrast, students who were taught concepts only through verbal or symbolic representations had lower performance and consequently greater difficulty in understanding the concepts. It is easy to conclude that the proper selection and use of multiple representations in the teaching of probability and statistical concepts leads to a better and deeper understanding of the concepts by students and maximizes the likelihood of their academic success.

REFERENCES

Ainsworth, S. (2006). DeFT: A conceptual framework for considering learning with multiple representations. *Learning and Instruction*, *16*(3), 183–198. DOI: 10.1016/j.learninstruc.2006.03.001

Ainsworth, S. E., Wood, D. J., & Bibby, P. A. (1997). Evaluating Principles for Multi-Representational Learning Environments. Paper presented at the 7th EARLI Conference. Athens.

Anasatsiadou, S., & Papadimitriou, I. (2003). Secondary mathematics teachers' attitudes towards statistics in Greece. 3° Mediterranean Conference in Mathematics Education, Athens, pp. 441-448.

Anastasiadou, S. (2002). Enhancing prospective teachers' knowledge of students' attitudes and representations towards statistics-the case of Greek students. In the Journal *'Scientia Paedacogika Experimentals'*. *XXXIX*, *1*, 103–127.

Anastasiadou, S. (2004a). Teacher's opinions about statistics and probability. In the journal *'Quaderni di Ricerca in Didattica'*, n_o 14. pp. 134-143. http://math.unipa.it/~grim/quaderno14.htm

Anastasiadou, S. (2004b). Perceptions-Attitudes-Conducts of the Greek Mathematicians for Statistics in Secondary Education, D. De Bock, M. Isoda, J. A. C. Cruz, A. Gagatsis, E Simmt (eds) 10th International Congress on Mathematical Education (ICME 10) Copenhagen, Demark, pp. 61-68.

Anastasiadou, S. (2005). Analysis of compound variables concerning students' attitudes towards statistics after the first contact with the lesson of statistics. In the journal *'Meditereeanean. Journal for Research in Mathematics Education*, *4*(1), 71–102.

Anastasiadou, S. (2005a). Survey of attitudes and beliefs toward Statistics in Greek Secondary Education International Conference on Mathematics Education, Svishtov, Bulgaria, pp. 125-133.

Anastasiadou, S. (2005b). Affective reactions and attitudes of the last class of Greek high school students towards statistics, In M Boosch (Eds) Proceedings 0r the Fourth Conference of the European Society for Research in Mathematics Education (CERME 4), Sant Feliu de Guíxols, Spain, 2005 in the http://cerme4.crm.es/Papers%20definitius/5/Anastasiadou%20Sofia.pdf

Anastasiadou, S. (2005c). Affective reactions and attitudes of the last class of Greek high school students towards statistics. Proceedings of CERME IV, European Research in Mathematics Education. Sant Feliu de Guíxols, Girona: CERME.

Anastasiadou, S. (2007a). The Role of representations in solving statistical problems and the translation ability of fifth and sixth grade students. *International Journal of Learning, 14*(10), 125–132. DOI: 10.18848/1447-9494/CGP/v14i10/45497

Anastasiadou, S. (2007b). Affective Responses and Attitudes of ninth grade Greek Students. *Proceedings of the 5th Mediterranean Conference in Mathematics Education*, Rodes, Greece pp. 169-178.

Anastasiadou, S. (2008a). Iconic Representations and Statistical Problem Solving: The Case of Greek Pupils. *International Journal of Learning, 15*(11), 301–310. DOI: 10.18848/1447-9494/CGP/v15i11/58699

Anastasiadou, S. (2008b). Representations and learning in statistics: a comparative study between Greek primary school students and immigrants. Eleventh International Congress on Mathematical Education, in its eleventh edition is to be held in Monterrey, Mexico, 2008 in the http://tsg.icme11.org/document/get/195

Anastasiadou, S. (2008c). The Effects of representational systems on the learning of statistics between Greek primary school students and immigrants. Proceeding of the 5th International Mathematics Education and Society Conference (MES 5) Albufeira, Portugal, pp.167-176.

Anastasiadou, S. (2009a). A structural equation model representing the hierarchical structure of translations among representations in Frequency concept. In Novotha, J., & Moraova, H. (Eds.), *International Symposium Elementary Mathematics Teaching (SEMT 09)* (pp. 57–64). Czech Repuplic., https://storage.cet.ac.il/cetnews/SEMT09 .PDF

Anastasiadou, S. (2009b). Greek students' ability in probability problem solving. In the Sixth Conference of the European Society for Research in Mathematics Education. Proceedings of CERME 6, Lyon France © INRP 2010 <www.inrp.fr/editions/cerme6> pp. 404-412. https://ife.ens-lyon.fr/editions/editions-electroniques/cerme6/working-group-3

Anastasiadou, S. (2010). Pre-service teachers' performance on the learning of probability distributions and the role of projects: A multilevel statistical analysis. In Regnier, Jean-Claude, Spagnolo, Filippo, Di Paola, Benedetto, & Gras, Regis, Eds, *5th International Conference Implicative Statistic Analysis*, Palermo, Italy *pp* 73-84.

Anastasiadou, S. (2019a). *Representations in statistics problem solving.ICERI2019, the* 12th annual International Conference of Education, Research and Innovation, Seville (Spain), ICERI2019, pp. 8993- 8996. DOI: 10.21125/iceri.2019.2159

Anastasiadou, S. (2019b). Internal representations toward external representations: The case of statistical concepts. 11th annual International Conference on Education and New Learning Technologies Palma de Mallorca, Spain, EDULEARN 19, pp. 6-9.

Anastasiadou, S. (2019c). Pre-service teachers' attitudes toward the use of different modes of representation in statistics teaching and learning. 11th annual International Conference on Education and New Learning Technologies Palma de Mallorca, Spain, EDULEARN 19. pp. 1-5.

Anastasiadou, S., Chadjipadelis, T., & Kofou, I. (2013a). Pre-school Education department to understand probability distributions. International Symposium Elementary Mathematics Teaching (SEMT '13), Prague, Czech Republic, In Jarmila Novotná and Hana Moraová ed., pp 67-75.

Anastasiadou, S., Chadjipadelis, T., & Kofou, I. (2013b). The role of projects in helping Greek university students from Pre-school Education department to understand probability distributions International Symposium Elementary Mathematics Teaching (SEMT '13). Prague, Czech Repuplic. International Symposium Elementary Maths Teaching. Proceedings: Tasks and tools in elementary mathematics, Prague [2013] σελ.67-74

Anastasiadou, S., & Chadjipantelis, T. (2008). The role of representations in the understanding of probabilities in tertiary education. Eleventh International Congress on Mathematical Education, Monterrey, Mexico, 2008 in the http://tsg.icme11.org/tsg/show/14

Anastasiadou, S., Elia, I., & Monoyiou, A. (2008). The effects of representations on the learning and understanding of statistics in Greek primary school. *Mediterranean Journal for Research in Mathematics Education*, *7*(1), 67–80.

Anastasiadou, S., & Gagatsis, A. (2005a). Attitudes vers la statistique des étudiants –futurs enseignants de l'école primaire grecque. In Gras, Regis, Spagnolo, Filippo, & David, Jerome Eds, *3nd International Conference Implicative Statistic Analysis*, Palermo, Italy, pp. 109-118.

Anastasiadou, S., & Gagatsis, A. (2005b). Believes and academic behaviors in statistics education. In Novotha, J. Eds, *International Symposium Elementary Mathematics Teaching SEMT 05 Prague*, pp. 59-65.

Anastasiadou, S., & Gagatsis, A. (2007). Exploring the effects of representation on the learning of statistics in Greek primary school. In the Fifth Conference of the European Society for Research in Mathematics Education (CERME 5), Nikosia, Cyprus pp. 90-100. http://ermeweb.free.fr/CERME5b/ και http://ermeweb.free.fr/CERME%205/WG1/1_Anastasiadou.pdf

Anastasiadou, S., & Gagatsis, A. (2017). Students' representations of linear motion. *Proceedings* of 9ème Colloque International sur Analyse Statistique Implicative. 9th International Meeting Statistical Implicative Analysis. Belfort – France, pp. 479-494.

Anastasiadou, S., & Papadimitriou, I. (2001a). *Another way to estimate the validity of a questionnaire for measurement of attitudes towards statistics. The case of Greek students. In the Journal 'Learning in Mathematics and Science and Educational Technology* (Vol. II). Intercollege Press.

Anastasiadou, S., & Papadimitriou, I. (2001b). *Another way to estimate the validity of a questionnaire for measurement of attitudes towards statistics. The case of Greek students. In the Journal 'Learning in Mathematics and Science and Educational Technology* (Vol. II). Intercollege Press.

Anastasiadou, S., & Pappa, A. (2019). *Greek pre-service teachers' perceptions, emotions and attitudes toward representations of physic concepts.* ICERI2019, the 12th annual International Conference of Education, Research and Innovation, Seville (Spain), ICERI2019, pp. 8987-8992. DOI: 10.21125/iceri.2019.2158

Chadjipantelis, T., & Anastasiadou, S. (2010). Pre-service teachers' understanding of probability Distributions: A Multilevel Statistical Analysis. International Conference on the Teaching of Statistics. ICOTS 8, Ljubljanja, Slovenia 2010. https://www.stat.auckland.ac.nz/~iase/publications.php?show=icots8

Christodoulou, T., Gagatsis, A., & Anastasiadou, S. (2017). Enquête internationale comparative: croyances et attitudes des élèves du secondaire inférieur relatives aux erreurs en mathématiques. *Proceedings* of 9ème Colloque International sur Analyse Statistique Implicative, Belfort – France, pp. 404-418.

DeWindt-King, A. M., & Goldin, G. A. (2003). Children's visual imagery: Aspects of cognitive representation in solving problems with fractions. *Mediterranean Journal for Research in Mathematics Education*, *2*(1), 1–42.

Dufur-Janvier, B., Bednarz, N., & Belanger, M. (1987). Pedagogical Considerations Concerning the Problem of representation. In Janvier, C. (Ed.), *Problem of Representation in the Teaching and Learning of Mathematics* (pp. 57–63). Lawrence Erlbaum.

Duval, R. (1987). The role of interpretation in learning mathematics. *Dimension, 2*, 56–74.

Duval, R. (2002). The cognitive analysis of problems of comprehension in the learning of mathematics. *Mediterranean Journal for Research in Mathematics Education, 1*(2), 1–16.

Gagatsis, A., Anastasiadou, S., & Mpora, E. (1997a). Errors made by Greek students of mathematics in probabilities, in the Journal '*Didactics of mathematics-Technology in Education',* pp. *47-66.*

Gagatsis, A., Anastasiadou, S., & Mpora, E. (1997b). Errori commess; da studenti greci di Matematica in questioni di probabilita, In the journal '. *Bollettino dei Docenti di Matematica, 35*, 87–102.

Goldin, G. A. (1998). Representational systems, learning, and problem solving in mathematics. *The Journal of Mathematical Behavior, 17*(2), 137–165. DOI: 10.1016/S0364-0213(99)80056-1

Janvier, C. (1987). Representation and understanding: The notion of functions as an example. In Janvier, C. (Ed.), *Problems of representation in the teaching and learning of mathematics* (pp. 67–72). Lawrence Erlbaum Associates.

Kaput, J. (1987). Representation and problem solving: methodological issues related to modelling. In Silver, E. A. (Ed.), *Teaching and Learning Mathematical Problem Solving: Multiple Research Perspectives* (pp. 381–398). Erlbaum.

Nicolaou, S., Gagatsis, A., Deliyianni, E., Panaoura, A., Elia, I., & Anastasiadou, S. (2017). Tracing the beliefs and self-efficacy beliefs of undergraduate economic sciences students: the case of the representations of functions. *Proceedings* of 9ème Colloque International sur Analyse Statistique Implicative. Belfort – France, pp. 461-478.

Petridou, A., Elia, I., Gagatsis, A., & Anastasiadou, S. (2017). L'appréhension perceptive des figures géométrique. *Proceedings* of 9ème Colloque International sur Analyse Statistique Implicative. Belfort – France, pp. 231-247.

Compilation of References

Aboaoja, F. A., Zainal, A., Ghaleb, F. A., Al-Rimy, B A S., Eisa, T. A. E., & Elnour, A. A. H. (2022). Malware detection issues, challenges, and future directions: A survey. *Applied Sciences (Basel, Switzerland)*, *12*(17), 8482. DOI: 10.3390/app12178482

Aigner, M. (2007). *Discrete Mathematics*. American Mathematical Society.

Ainsworth, S. E., Wood, D. J., & Bibby, P. A. (1997). Evaluating Principles for Multi-Representational Learning Environments. Paper presented at the 7th EARLI Conference. Athens.

Ainsworth, S. (2006). DeFT: A conceptual framework for considering learning with multiple representations. *Learning and Instruction*, *16*(3), 183–198. DOI: 10.1016/j.learninstruc.2006.03.001

Ainsworth, S. (2020). The educational value of multiple representations: Lessons from cognitive science. *Learning and Instruction*, *65*, 101–120.

Akala, B. M. (2021). Revisiting education reform in Kenya: A case of Competency-Based Curriculum (CBC). *Social Sciences & Humanities Open*, *3*(1), 100129.

Alalwani, M H J, ALhamad A M. (2021). A framework for investigating the effect of unlimited improvement on quality of education. *International Journal of Academic and Applied Research*, *5*(6), 1–10.

Alam, A. (2023). “Improving Learning Outcomes through Predictive Analytics: Enhancing Teaching and Learning with Educational Data Mining.” 2023 7th International Conference on Intelligent. DOI: 10.1109/ICICCS56967.2023.10142392

Alcover, R., Chirivella, V., & Vazquez, E. (2024). A proposal to speed up the learning of statistics using generative AI. *Proceedings of the 17th Annual International Conference of Education, Research and Innovation*, 3697-3701, Seville, Spain. DOI: 10.21125/iceri.2024.0944

Alenezi, M. (2023). Digital learning and digital institution in higher education. *Education Sciences*, *13*(4), 1–10. DOI: 10.3390/educsci13010088

Aliaga, M., Cuff, C., Garfield, J., & Lock, R. (2005). *GAISE Report: Guidelines for Assessment and Instruction in Statistics Education*. American Statistical Association.

Alon, N., & Spencer, J. H. (2000). *The Probabilistic Method* (2nd ed.). Wiley. DOI: 10.1002/0471722154

Anand, R. (2007). Digital visualizations for self-directed learning in Indian universities. *Journal of Higher Education Research*, *25*(1), 33–47.

Anasatsiadou, S., & Papadimitriou, I. (2003). Secondary mathematics teachers' attitudes towards statistics in Greece. 3° Mediterranean Conference in Mathematics Education, Athens, pp. 441-448.

Anastasiadou, S. (2004). Teacher's opinions about statistics and probability. In the journal 'Quaderni di Ricerca in Didattica', no 14. pp. 134-143. http://math.unipa .it/~grim/quaderno14.htm

Anastasiadou, S. (2004a). Teacher's opinions about statistics and probability. In the journal *'Quaderni di Ricerca in Didattica'*, n_o 14. pp. 134-143. http://math.unipa .it/~grim/quaderno14.htm

Anastasiadou, S. (2004b). Perceptions-Attitudes-Conducts of the Greek Mathematicians for Statistics in Secondary Education, D. De Bock, M. Isoda, J. A. C. Cruz, A. Gagatsis, E Simmt (eds) 10th International Congress on Mathematical Education (ICME 10) Copenhagen, Demark, pp. 61-68.

Anastasiadou, S. (2005). Affective reactions and attitudes of the last class of Greek high school students towards statistics, In M Boosch (Eds) *Proceedings Or the Fourth Conference of the European Society for Research in Mathematics Education* (CERME 4), Sant Feliu de Guíxols, Spain, 2005 in the http://cerme4.crm.es/Papers %20definitius/5/Anastasiadou%20Sofia.pdf

Anastasiadou, S. (2005). Affective reactions and attitudes of the last class of Greek high school students towards statistics, In M Boosch (Eds) *Proceedings of the Fourth Conference of the European Society for Research in Mathematics Education* (CERME 4), Sant Feliu de Guíxols, Spain.

Anastasiadou, S. (2005a). Survey of attitudes and beliefs toward Statistics in Greek Secondary Education International Conference on Mathematics Education, Svishtov, Bulgaria, pp. 125-133.

Anastasiadou, S. (2005b). Affective reactions and attitudes of the last class of Greek high school students towards statistics, In M Boosch (Eds) Proceedings Or the Fourth Conference of the European Society for Research in Mathematics Education (CERME 4), Sant Feliu de Guíxols, Spain, 2005 in the http://cerme4.crm.es/Papers %20definitius/5/Anastasiadou%20Sofia.pdf

Anastasiadou, S. (2005c). Affective reactions and attitudes of the last class of Greek high school students towards statistics. Proceedings of CERME IV, European Research in Mathematics Education. Sant Feliu de Guíxols, Girona: CERME.

Anastasiadou, S. (2005c). Survey of attitudes and beliefs toward Statistics in Greek Secondary Education International Conference on Mathematics Education, Svishtov, Bulgaria, pp. 125-133.

Anastasiadou, S. (2008b). Representations and learning in statistics: a comparative study between Greek primary school students and immigrants. Eleventh International Congress on Mathematical Education, in its eleventh edition is to be held in Monterrey, Mexico, 2008 in the http://tsg.icme11.org/document/get/195

Anastasiadou, S. (2008c). The Effects of representational systems on the learning of statistics between Greek primary school students and immigrants. Proceeding of the 5th International Mathematics Education and Society Conference (MES 5) Albufeira, Portugal, pp.167-176.

Anastasiadou, S. (2009a). Greek students' ability in probability problem solving. In the Sixth Conference of the European Society for Research in Mathematics Education. Proceedings of CERME 6, Lyon France © INRP 2010 <www.inrp.fr/editions/cerme6> pp. 404-412. https://ife.ens-lyon.fr/editions/editions-electroniques/cerme6/working-group-3

Anastasiadou, S. (2009a). Greek students' ability in probability problem solving. In the Sixth Conference of the European Society for Research in Mathematics Education. Proceedings of CERME 6, Lyon France © INRP 2010, pp. 404-412. https://ife.ens-lyon.fr/editions/editions-electroniques/cerme6/working-group-3

Anastasiadou, S. (2009b). Greek students' ability in probability problem solving. In the Sixth Conference of the European Society for Research in Mathematics Education. Proceedings of CERME 6, Lyon France © INRP 2010 <www.inrp.fr/editions/cerme6> pp. 404-412. https://ife.ens-lyon.fr/editions/editions-electroniques/cerme6/working-group-3

Anastasiadou, S. (2016). *Evaluation of the implementation of TQM principles in tertiary education using the EFQM excellence model: Research in educational departments of Greek universities* [Master's thesis, Greek Open University]. Patra, Greece.

Anastasiadou, S. (2016). Principal Component Analysis vs. Implicative Statistical Analysis: The case of probability distributions' understanding. Global Journal of Psychology Research. Global Journal on Humanites & Social Sciences. [Online]. 00, pp 00-00. Available from: http://sproc.org/ojs/index.php/pntsbs

Anastasiadou, S. (2018). Comparison of multivariate methods in group/cluster identification. Dissertation thesis of the degree of MSc. in Research Methodology in Biomedicine, Biostatistics and Clinical Bioinformatics, Faculty of Medicine, University of Thessaly.

Anastasiadou, S. (2018). Comparison of multivariate methods in group/cluster identification. Dissertation thesis of the degree of MSc. in Research Methodology in Biomedicine, Biostatistics and Clinical Bioinformatics, Fuculty of Medicine, University of Thessaly.

Anastasiadou, S. (2019a). Pre-service teachers' attitudes toward the use of different modes of representation in statistics teaching and learning. 11th annual International Conference on Education and New Learning Technologies Palma de Mallorca, Spain, EDULEARN 19. pp. 1-5.

Anastasiadou, S. (2019a). *Representations in statistics problem solving*. ICERI2019, Proceedings of the 12th annual International Conference of Education, Research and Innovation, Seville (Spain), ICERI2019, pp. 8993- 8996. DOI: 10.21125/iceri.2019.2159

Anastasiadou, S. (2019b). Internal representations toward external representations: The case of statistical concepts. 11th annual International Conference on Education and New Learning Technologies Palma de Mallorca, Spain, EDULEARN 19, pp. 6-9.

Anastasiadou, S. (2019b). *Internal representations toward external representations: The case of statistical concepts*. 11th annual International Conference on Education and New Learning Technologies, Palma de Mallorca, Spain, EDULEARN 19, pp. 6-9. DOI: 10.21125/edulearn.2019.0004

Anastasiadou, S. (2019b). Pre-service teachers' attitudes toward the use of different modes of representation in statistics teaching and learning. 11th annual International Conference on Education and New Learning Technologies Palma de Mallorca, Spain, EDULEARN 19. pp. 1-5.

Anastasiadou, S. (2019c). Pre-service teachers' attitudes toward the use of different modes of representation in statistics teaching and learning. 11th annual International Conference on Education and New Learning Technologies Palma de Mallorca, Spain, EDULEARN 19. pp. 1-5.

Anastasiadou, S. Zirinoglou, P. (2025). L'Analyse Factorielle des Correspondances Reveal Students' Attitudes Toward Entrepreneurship Source Title: Real-World Tools and Scenarios for Entrepreneurship Exploration, (2025) 79-100. .DOI: 10.4018/979-8-3693-3100-2.ch004

Anastasiadou, S., & Chadjipantelis, T. (2008). The role of representations in the understanding of probabilities in tertiary education. Eleventh International Congress on Mathematical Education, Monterrey, Mexico, 2008 in the http://tsg.icme11.org/tsg/show/14

Anastasiadou, S., & Gagatsis, A. (2007). Exploring the effects of representation on the learning of statistics in Greek primary school. In the Fifth Conference of the European Society for Research in Mathematics Education (CERME 5), Nikosia, Cyprus pp. 90-100. http://ermeweb.free.fr/CERME5b/ και http://ermeweb.free.fr/CERME%205/WG1/1_Anastasiadou.pdf

Anastasiadou, S., & Gagatsis, A. (2017). Students' representations of linear motion. *Proceedings* of 9ème Colloque International sur Analyse Statistique Implicative. 9th International Meeting Statistical Implicative Analysis. Belfort – France, pp. 479-494.

Anastasiadou, S., & Masouras, A. (2023). Exploring determinants regarding Visit and Revisit intentions considering Greece as a Medical Tourism destination. 10th International Conference on Strategic Innovative Marketing and Tourism. Zakynthos, Ionian Islands, Greece. In A. Kavoura et al. (eds.), Strategic Innovative Marketing and Tourism, Springer Proceedings in Business and Economics, (2023), 101-109. DOI: 10.1007/978-3-031-51038-0_12

Anastasiadou, S., & Pappa, A. (2019). *Greek pre-service teachers' perceptions, emotions and attitudes toward representations of physic concepts.* ICERI2019, the 12th annual International Conference of Education, Research and Innovation, Seville (Spain), ICERI2019, pp. 8987-8992. DOI: 10.21125/iceri.2019.2158

Anastasiadou, S., & Taraza, E. (2019a). Total quality management: implementation of the six sigma methodology for improving quality in higher education, ICERI2019, the 12th annual International Conference of Education, Research and Innovation, Seville (Spain), ICERI2019, pp. 9533-9537.

Anastasiadou, S., & Taraza, E. (2019b). Total Quality Management: Statistical Processes and Quality Organizations' Measurement Tools. ICERI2019, the 12th annual International Conference of Education, Research and Innovation will be held in Seville (Spain), ICERI2019.

Anastasiadou, S., & Taraza, E. (2020). Six Sigma in Tertiary Education: A Win of Change regarding Quality Improvement in Education, Proceedings of of 14th annual International Technology, Education and Development Conference (INTED2020), Valencia, Spain, pp. 9595-9601. DOI: 10.21125/inted.2020.0167

Anastasiadou, S., Chadjipadelis, T., & Kofou, I. (2013). Pre-school Education department to understand probability distributions. International Symposium Elementary Mathematics Teaching (SEMT '13), Prague, Czech Republic, In Jarmila Novotná and Hana Moraová ed., pp 67-75.

Anastasiadou, S., Chadjipadelis, T., & Kofou, I. (2013). The role of projects in helping Greek university students from Pre-school Education department to understand probability distributions International Symposium Elementary Mathematics Teaching (SEMT '13). Prague, Czech Repuplic. International Symposium Elementary Maths Teaching. Proceedings: Tasks and tools in elementary mathematics, Prague [2013] σελ.67-74

Anastasiadou, S., Chadjipadelis, T., & Kofou, I. (2013a). Pre-school Education department to understand probability distributions. International Symposium Elementary Mathematics Teaching (SEMT '13), Prague, Czech Republic, In Jarmila Novotná and Hana Moraová (Eds). pp 67-75.

Anastasiadou, S., Chadjipadelis, T., & Kofou, I. (2013a). Pre-school Education department to understand probability distributions. International Symposium Elementary Mathematics Teaching (SEMT '13), Prague, Czech Republic, In Jarmila Novotná and Hana Moraová ed., pp 67-75.

Anastasiadou, S., Chadjipadelis, T., & Kofou, I. (2013a). The role of projects in helping Greek university students from Pre-school Education department to understand probability distributions International Symposium Elementary Mathematics Teaching (SEMT '13). Prague, Czech Repuplic. International Symposium Elementary Maths Teaching. Proceedings: Tasks and tools in elementary mathematics, Prague [2013] σελ.67-74.

Anastasiadou, S., Chadjipadelis, T., & Kofou, I. (2013b). The role of projects in helping Greek university students from Pre-school Education department to understand probability distributions International Symposium Elementary Mathematics Teaching (SEMT '13). Prague, Czech Repuplic. International Symposium Elementary Maths Teaching. Proceedings: Tasks and tools in elementary mathematics, Prague [2013] σελ.67-74

Anastasiadou, S., Papalexandris, S. & Konteos, G. (2022). The Effect of Emotional Intelligence on Perceived e-Service Quality, Consumers' Perceived Value, Purchase, Loyalty Intentions and Satisfaction, Haicta. CEUR Workshop Proceedings (CEUR-WS.org), 151-161.

Anastasiadou, S., Zirinoglou, P., Seremeti, L., Masouras, A., Komodromos, M. & Anastasiadis, L. (2024). Statistical Implicative Analysis: A Distinguished Statistical Methodology in Evaluating Greek Students' Attitudes Toward the Use of AI Source Title: Real-World Tools and Scenarios for Entrepreneurship Exploration, (2024), 79-100, .DOI: 10.4018/979-8-3693-3100-2.ch006

Anastasiadou, SD. Anastasiadis, L. Florou, G. & Valsamidis, S. (2022). Greek students' Environmental Attitude based on the Theory of Ecological Attitude. HA-ICTA, (2022), 222-231.

Anastasiadou, S. (2002). Enhancing prospective teachers' knowledge of students' attitudes and representations towards statistics-the case of Greek students. In the Journal 'Scientia Paedacogika Experimentals'. *XXXIX*, *1*, 103–127.

Anastasiadou, S. (2005). Analysis of compound variables concerning students' attitudes towards statistics after the first contact with the lesson of statistics. In the journal *'Meditereeanean. Journal for Research in Mathematics Education*, *4*(1), 71–102.

Anastasiadou, S. (2007). Affective Responses and Attitudes of ninth grade Greek Students. *Proceedings of the 5th Mediterranean Conference in Mathematics Education*, Rodes, Greece pp. 169-178.

Anastasiadou, S. (2007). T*he Role of representations in solving statistical problems and the translation ability of fifth and sixth grade students.Proceedings of the International Journal of Learning*, *14*(10), 125–132.

Anastasiadou, S. (2007a). The Role of representations in solving statistical problems and the translation ability of fifth and sixth grade students. *International Journal of Learning*, *14*(10), 125–132. DOI: 10.18848/1447-9494/CGP/v14i10/45497

Anastasiadou, S. (2008). Iconic Representations and Statistical Problem Solving: The Case of Greek Pupils. *International Journal of Learning*, *15*(11), 301–310. DOI: 10.18848/1447-9494/CGP/v15i11/58699

Anastasiadou, S. (2009a). A structural equation model representing the hierarchical structure of translations among representations in Frequency concept. In Novotha, J., & Moraova, H. (Eds.), *International Symposium Elementary Mathematics Teaching (SEMT 09)* (pp. 57–64). Czech Repuplic., https://storage.cet.ac.il/cetnews/SEMT09 .PDF

Anastasiadou, S. (2009b). Detection of Factors Influencing Greek Technological Institute Students' Choice of Study Department with the Aid of Correspondence Factor Analysis and Path Analysis. *Journal of the World Universities Forum*, *2*(4), 131–142. DOI: 10.18848/1835-2030/CGP/v02i04/56606

Anastasiadou, S. (2010). *Pre-service teachers' performance on the learning of probability distributions and the role of projects: A multilevel statistical analysis.* In Regnier, Jean-Claude, Spagnolo, Filippo, Di Paola, Benedetto, & Gras, Regis, (Eds), *Proceedings of the 5th International Conference Implicative Statistic Analysis*, Palermo, Italy pp 73-84.

Anastasiadou, S. (2010). Pre-service teachers' performance on the learning of probability distributions and the role of projects: A multilevel statistical analysis. In Regnier, Jean-Claude, Spagnolo, Filippo, Di Paola, Benedetto, & Gras, Regis, (Eds). *5th International Conference Implicative Statistic Analysis*, Palermo, Italy pp 73-84.

Anastasiadou, S. (2013). Developing and Evaluating a Structural Equation Model Measuring Leadership Changes in a Lifelong Learning World. *The International Journal of Educational Organization and Leadership*, *19*(2), 1–17. DOI: 10.18848/2329-1656/CGP/v19i02/48520

Anastasiadou, S. (2014). A Structural Equation Model Describes Factors Affecting Greek Students' Consumer behavior. *Procedia Economics and Finance*, *9*(1), 402–406. DOI: 10.1016/S2212-5671(14)00041-0

Anastasiadou, S. (2019c). Comparison of contemporary advanced statistical methods regarding construct validity evaluation of TEIque-SF instrument: Statistical Implicative Analysis vs. Principal Components Analysis. *9ème Colloque International sur Analyse Statistique Implicative (ASI 10)*. Belfort – France. pp. 148-163.

Anastasiadou, S. (2019d). Comparison of contemporary advanced statistical methods regarding construct validity evaluation of TEIque-SF instrument: Statistical implicative analysis vs. principal components analysis. In *9ème Colloque International sur Analyse Statistique Implicative (ASI 10)* (pp. 148–163). Belfort, France.

Anastasiadou, S. (2023). Psychometric Testing Procedures for Attitudes Toward Reproductive Tourism and Cross Border Reproductive Care (ATCBRC). In Kavoura, A., Borges-Tiago, T., & Tiago, F. (Eds.), *Strategic Innovative Marketing and Tourism. ICSIMAT 2023. Springer Proceedings in Business and Economics*. Springer., DOI: 10.1007/978-3-031-51038-0_14

Anastasiadou, S. D., Batiou, V., & Balkanos, E. (2014). Occupation mobility in Greece: A completely new area. Global Journal of Business. *Economics and Management*, *4*(2), 73–80.

Anastasiadou, S., & Anastasiadis, L. (2019). Quality Assurance in Education in the Light of the Effectiveness of Transformational School Leadership. In Sykianakis, N., Polychronidou, P., & Karasavvoglou, A. (Eds.), *Economic and Financial Challenges for Eastern Europe. Springer Proceedings in Business and Economics* (pp. 323–344). Springer., DOI: 10.1007/978-3-030-12169-3_21

Anastasiadou, S., Anastasiadis, L., Angeletos, T., & Vandikas, J. (2010b). A Multidimensional Statistical Analysis of Students' Attitudes toward Physics. *The International Journal of Diversity in Organisations, Communities and Nations*, *16*(5), 341–356. DOI: 10.18848/1447-9532/CGP/v10i01/39794

Anastasiadou, S., Anastasiadis, L., Vandikas, J., & Angeletos, T. (2010a). Implicative statistical analysis and Principal Components Analysis in recording students' attitudes toward electronics and electrical constructions subject. *International Journal of Technology, Knowledge and Society*, *16*(5), 341–356.

Anastasiadou, S., Elia, I., & Monoyiou, A. (2008). The effects of representations on the learning and understanding of statistics in Greek primary school. *Mediterranean Journal for Research in Mathematics Education*, *7*(1), 67–80.

Anastasiadou, S., & Gagatsis, A. (2005a). Attitudes vers la statistique des étudiants –futurs enseignants de l'école primaire grecque. In Gras, Regis, Spagnolo, Filippo, & David, Jerome Eds, *3rd International Conference Implicative Statistic Analysis*, Palermo, Italy, pp. 109-118.

Anastasiadou, S., & Gagatsis, A. (2005b). Believes and academic behaviors in statistics education. In Novotha, J. Eds, *International Symposium Elementary Mathematics Teaching SEMT 05 Prague*, pp. 59-65.

Anastasiadou, S., & Karakos, A. (2011). The beliefs of electrical and computer engineering students´ regarding computer programming. *International Journal of Technology, Knowledge and Society*, *7*(1), 37–52. DOI: 10.18848/1832-3669/CGP/v07i01/56170

Anastasiadou, S., & Loukas, D. (2009). Greek pre-service teachers' cognitive abilities in understanding the concept of frequency: A multilevel statistical analysis. *International Journal of Learning, 16*(5), 189–202. DOI: 10.18848/1447-9494/CGP/v16i05/46268

Anastasiadou, S., Masouras, A., & Papademetriou, C. (2023). Attitudes toward Reproductive Tourism and Cross border reproductive care (CBRC): Legal, Economic, Ethical issues and dilemmas, possibilities and limitations. *Proceedings of the 6th International Conference on Tourism Research, ICTR 2023*, 7-23, Pafos, Cyprus. DOI: 10.34190/ictr.6.1.1189

Anastasiadou, S., Masouras, A., & Papademetriou, C. (2023). Attitudes toward Repro-ductive Tourism and Cross border reproductive care (CBRC): Legal, Economic, Eth-ical issues and dilemmas, possibilities and limitations. *Proceedings of the 6th Inter-national Conference on Tourism Research, ICTR 2023*, Pafos, Cyprus, pp. 17-23.

Anastasiadou, S., & Papadimitriou, I. (2001a). *Another way to estimate the validity of a questionnaire for measurement of attitudes towards statistics. The case of Greek students. In the Journal 'Learning in Mathematics and Science and Educational Technology* (Vol. II). Intercollege Press.

Angelis, D., Sofos, F., & Karakasidis, T. E. (2023). Artificial intelligence in physical sciences: Symbolic regression trends and perspectives. *Archives of Computational Methods in Engineering, 30*(6), 1–15. DOI: 10.1007/s11831-023-09922-z PMID: 37359747

Auzmendi, E. (1992). *Las actitudes hacia la matemática-estadística en las enseñanzas medias y universitarios*. Mensajero.

Bagha, K. N. (2011). A short introduction to semantics. *Journal of Language Teaching and Research, 2*(6), 1411–1419. DOI: 10.4304/jltr.2.6.1411-1419

Baron, M. (2014). *Probability and Statistics for Computer Science*. CRC Press.

Basar, Z. M., Mansor, A. N., & Jamaludin, K. A. (2021). The Effectiveness and Challenges of Online Learning for Secondary School Students–A Case Study. *Journal of University Education, 2*(1), 1–15.

Batanero, C., & Alvarez-Arroyo, R. (2024). Teaching and learning of probability. *ZDM Mathematics Education, 56*(1), 1–13. DOI: 10.1007/s11858-023-01511-5

Beichner, R. J. (1994). Graph interpretation and its role in statistical reasoning in STEM education. *Journal of Research in Science Teaching, 31*(9), 839–855.

Ben-Zvi, D., & Garfield, J. (2019). *Developing students' statistical reasoning: Connecting research and teaching practice*. Springer.

Beranger, B., Lin, H., & Sisson, S. (2023). New Models for Symbolic Data Analysis. *Advances in Data Analysis and Classification*, *17*(1), 1–18.

Berthet, V. (2022). The impact of cognitive biases on professionals' decision-making: A review of four occupational areas. *Frontiers in Psychology*, *13*, 1024941. DOI: 10.3389/fpsyg.2021.802439 PMID: 35058862

Bindra, K., & Mishra, A. (2023). A detailed study of clustering algorithms. *Proceedings of the 6th International Conference on Reliability: Infocom Technologies and Optimization (Trends, and Future Directions)*, Noida, India, 371-376. DOI: 10.1109/ICRITO.2017.8342454

Black, K. (2023). *Business Statistics for contemporary decision making*. Wiley. ISBN: 978-1-119-9054-8.

Boaler, J. (2016). *Mathematical mindsets: Unleashing students' potential through creative math, inspiring messages and innovative teaching*. Jossey-Bass.

Bodin, A., Courtourier, R., & Gras, R. (2000). *CHIC: Classification Hiérarchique Implicative et Cohesitive-Version sous Windows – CHIC 1.2*. Association pour le Recherche en Didactique des Mathématiques.

Boyd, R. L., & Schwartz, H. A. (2021). Natural language analysis and the psychology of verbal behavior: The past, present, and future states of the field. *Journal of Language and Social Psychology*, *40*(1), 1–15. DOI: 10.1177/0261927X20967028 PMID: 34413563

Brown, G. D., Largey, A., & McMullan, C. (2021). The impact of gender on risk perception: Implications for EU member states' national risk assessment processes. *International Journal of Disaster Risk Reduction*, *55*, 1–12. DOI: 10.1016/j.ijdrr.2021.102452

Budig, M. J., & England, P. (2001). The wage penalty for motherhood. *American Sociological Review*, *66*(2), 204–225. DOI: 10.1177/000312240106600203

Burch, M., Ten Brinke, K. B., Castella, A., Peters, G. K. S., Shteriyanov, V., & Vlasvinkel, R. (2021). Dynamic graph exploration by interactively linked node-link diagrams and matrix visualizations. *Visual Computing for Industry, Biomedicine, and Art*, *4*(1), 23. DOI: 10.1186/s42492-021-00088-8 PMID: 34491465

Burgos, M., Batanero, C., & Godino, J. D. (2021). Algebraization levels in the study of probability. *Mathematics*, *9*(11), 1342. DOI: 10.3390/math10010091

Cai, S., Liu, E., Shen, Y., Liu, C., Li, S., & Shen, Y. (2023). Probability learning in mathematics using augmented reality: Impact on students' learning gains and attitudes. *International Journal of ILEs in Education, 11*(1), 1–18. DOI: 10.4324/9781003457121-3

Calonge, D. S., Smail, L., & Kamalov, F. (2023). Enough of the chit-chat: A comparative analysis of four AI chatbots for calculus and statistics. *Journal of Applied Learning & Teaching, 6*(2). Advance online publication. DOI: 10.37074/jalt.2023.6.2.22

Cascão, A., Quelhas, A. P., & Cunha, A. M. (2023). Heuristics and cognitive biases in the housing investment market. *Journal of Housing Markets and Analysis, 12*(1), 1–18. DOI: 10.1108/IJHMA-05-2022-0073

Chadjipantelis, T., & Anastasiadou, S. (2010). Pre-service teachers' understanding of probability Distributions: A Multilevel Statistical Analysis. International Conference on the Teaching of Statistics. ICOTS 8, Ljubljanja, Slovenia 2010. https://www.stat.auckland.ac.nz/~iase/publications.php?show=icots8

Chance, B. L. (2017). Components of statistical thinking and implications for instruction and assessment. *Journal of Statistics Education : An International Journal on the Teaching and Learning of Statistics, 10*(3). Advance online publication. DOI: 10.1080/10691898.2002.11910677

Chan, K. (2015). *Cognitive effects of augmented PowerPoint presentations on statistical understanding in Indian classrooms*. Educational Studies Repository.

Chen, S., & Yen, S. (2015). Multimedia learning principles for reducing cognitive load and improving retention. *Computers in Human Behavior, 50*(3), 522–531.

Cheung, R. W., Hartley, C., & Monaghan, P. (2022). Receptive and expressive language ability differentially support symbolic understanding over time: Picture comprehension in late talking and typically developing children. *Journal of Experimental Child Psychology, 223*, 105564. DOI: 10.1016/j.jecp.2021.105305 PMID: 34653634

Chick, H. L., & Pierce, R. (2019). Teachers' beliefs and practices regarding multiple representations in statistics education. *International Journal of Mathematics Education, 51*(4), 56–74.

Christodoulou, T., Gagatsis, A., & Anastasiadou, S. (2017). Enquête internationale comparative: croyances et attitudes des élèves du secondaire inférieur relatives aux erreurs en mathématiques. *Proceedings* of 9ème Colloque International sur Analyse Statistique Implicative, Belfort – France, pp. 404-418.

Coakes, S., & Steed, L. (1999). *SPSS Analysis without Anguish.* John Willey & Sons.

Courturier, R. (2008). CHIC: Cohesive Hierarchical Implicative Classification. In Gras, R., Suzuki, E., Guillet, F., & Spagnolo, F. (Eds.), *Statistical Implicative Analysis* (pp. 1–13). Springer Verlag Berlin Heidelberg. DOI: 10.1007/978-3-540-78983-3_2

Coutourier, Raphael., & Gras, Rgis. (2005). CHIC: Traitement de données avec L; analyse implicative Extraction et Gestion des Connaissances, Volume II, RMTI, Cepadues, Paris, pp. 679-684.

Couturier, R. & Pazmiño, R. (2015). Statistical implicative analysis for educational data sets: 2 analysis with RCHI, XVIII Congreso Internacional EDUTEC "Educación y Tecnología desde una visión Transformadora", pp. 1-11

Couturier, R. (2008). *CHIC: Cohensive Hierarchical Implicative Classification. Studies in Computational Intelligence (SCI).* Springer-Verlag Berlin Heidelberg.

Crenshaw, K. (1991). Mapping the margins: Intersectionality, identity politics, and violence against women of color. *Stanford Law Review, 43*(6), 1241–1299. DOI: 10.2307/1229039

Curth, A., & Van der Schaar, M. (2021). On inductive biases for heterogeneous treatment effect estimation. *Advances in Neural Information Processing Systems.*

Das, R., & Soylu, M. (2023). A key review on graph data science: The power of graphs in scientific studies. *Chemometrics and Intelligent Laboratory Systems, 229*, 104654. DOI: 10.1016/j.chemolab.2023.104896

Das, S., & Roy, P. (2014). Cultural considerations in designing visual tools for Indian classrooms. *Asian Educational Research, 19*(4), 200–212.

Demszky D, Liu J, Hill H C. (2024). Can automated feedback improve teachers' uptake of student ideas? Evidence from a randomized controlled trial in a large-scale online course. Journal of Educational Data Mining and Learning Analytics.

Demszky, D., Liu, J., Hill, H. C., Jurafsky, D., & Piech, C. (2024). Can automated feedback improve teachers' uptake of student ideas? Evidence from a randomized controlled trial in a large-scale online course. *Educational Evaluation and Policy Analysis, 46*(1), 1–15. DOI: 10.3102/01623737231169270

DeWindt-King, A. M., & Goldin, G. A. (2003). Children's visual imagery: Aspects of cognitive representation in solving problems with fractions. *Mediterranean Journal for Research in Mathematics Education, 2*(1), 1–42.

Dhami, M. K., & Mandel, D. R. (2021). Words or numbers? Communicating probability in intelligence analysis. *The American Psychologist, 76*(1), 1–13. DOI: 10.1037/amp0000637 PMID: 32700939

Dufur-Janvier, B., Bednarz, N., & Belanger, M. (1987). Pedagogical Considerations Concerning the Problem of representation. In Janvier, C. (Ed.), *Problem of Representation in the Teaching and Learning of Mathematics* (pp. 57–63). Lawrence Erlbaum.

Duval R (2006) A cognitive analysis of problems of com447 prehension in a learning of mathematics. Educ Stud, 448 Math 61:103–131

Duval, R. (1987). The role of interpretation in learning mathematics. *Dimension, 2*, 56–74.

Duval, R. (2002). The cognitive analysis of problems of comprehension in the learning of mathematics. *Mediterranean Journal for Research in Mathematics Education, 1*(2), 1–16.

Eagly, A. H., & Carli, L. L. (2007). *Through the labyrinth: The truth about how women become leaders*. Harvard Business Press.

Eden, C.A., Chison, O.N. and Ademiyi, I.S. (2024). Integrating AI in education: Opportunities, challenges, and ethical considerations. *Magna Scientia Advanced Research and Reviews*, 10(02), 06-013. DOI: 10.30574/msarr.2024.10.2.0039

Efthymiou, A., Rudinac, S., Kackovic, M., & Worring, M. (2021). "Graph neural networks for knowledge enhanced visual representation of paintings." arXiv preprint arXiv:2109.07633. DOI: 10.1145/3474085.3475586

Eilam, B. (2012). The role of visual literacy in integrating disciplinary literacy and cognitive comprehension. *Educational Researcher, 41*(2), 70–78.

Ellis, A. R., & Slade, E. (2023). A new era of learning: Considerations for ChatGPT as a tool to enhance statistics and data science education. *Journal of Statistics and Data Science Education : An Official Journal of the of the American Statistical Association, 31*(2), 128–133. DOI: 10.1080/26939169.2023.2223609

Ericsson, L., Gouk, H., Loy, C. C., & Hospedales, T. M. (2022). Self-supervised representation learning: Introduction, advances, and challenges. *IEEE Signal Processing Magazine, 39*(3), 42–62. DOI: 10.1109/MSP.2021.3134634

Estrella, S., Mendez-Reina, M., Olfos, R., & Aguilera, J. (2022). Early statistics in kindergarten: Analysis of an educator's pedagogical content knowledge in lessons promoting informal inferential reasoning. *International Journal for Lesson and Learning Studies, 11*(2), 1–15. DOI: 10.1108/IJLLS-07-2021-0061

Farahani, A. M., Adibi, P., Ehsani, M. S., & Hutter, H. P. (2023). Automatic Chart Understanding: A Review. *IEEE Transactions on Knowledge and Data Engineering*, *35*(1), 261–274.

Ferster B. (2023). Interactive visualization: Insight through inquiry.

Ferster, B. (2023). *Interactive visualization: Insight through inquiry*. CRC Press.

Fotiadis, Th., & Anastasiadou, S. (2018a). Contemporary advanced statistical methods for the science of marketing: Implicative Statistical Analysis vs Principal Components Analysis. Proceeding of the ICME 2018- 1st International Conference on Marketing and Entrepreneuship. Cyprus. https://hephaestus.nup.ac.cy/handle/11728/11393

Fotiadis, Th., & Anastasiadou, S. (2018a). Contemporary advanced statistical methods for the science of marketing: Principal Components Analysis vs Analysee Factorielle des Correspondances. Proceeding of the ICME 2018- 1st International Conference on Marketing and Entrepreneuship. Cyprus. https://hephaestus.nup.ac.cy/handle/11728/11392

Fotiadis, Th., & Anastasiadou, S. (2018b). Contemporary advanced statistical methods for the science of marketing: Implicative Statistical Analysis vs Principal Components Analysis. Proceeding of the ICME 2018- 1st International Conference on Marketing and Entrepreneuship. Cyprus. https://hephaestus.nup.ac.cy/handle/11728/11393

Fotiadis, Th., & Anastasiadou, S. (2018b). Contemporary advanced statistical methods for the science of marketing: Principal Components Analysis vs Analyse Factorielle des Correspondances. In *Proceedings of the ICME 2018 – 1st International Conference on Marketing and Entrepreneurship*. Cyprus. Retrieved from https://hephaestus.nup.ac.cy/handle/11728/11392

Franconeri, S. L., Padilla, L. M., & Shah, P. (2021). The Science of Visual Data Communication: What Works. *Science in Context*, *34*(2), 143–155. PMID: 34907835

Franklin, C., Kader, G., & Mewborn, D. (2020). Statistical thinking in the 21st century classroom. *Teaching Statistics*, *42*(3), 103–112.

Friedman, H. H. (2023). Cognitive biases and their influence on critical thinking and scientific reasoning: A practical guide for students and teachers.

Fries, L., Son, J. Y., Givvin, K. B., & Stigler, J. W. (2021). Practicing connections: A framework to guide instructional design for developing understanding in complex domains. *Educational Psychology Review*, *33*(2), 739–762. DOI: 10.1007/s10648-020-09561-x

Furlong, M., & Eliasmith, C. (2022). "Fractional binding in vector symbolic architectures as quasi-probability statements." *Proceedings of the Annual Meeting of the Cognitive Science Society*, 44, 1-13.

Gagatsis, A., Anastasiadou, S., & Mpora, E. (1997a). Errors made by Greek students of mathematics in probabilities, in the Journal '*Didactics of mathematics-Technology in Education'*, pp. *47-66.*

Gagatsis, A., Anastasiadou, S., & Mpora, E. (1997b). Errori commess; da studenti greci di Matematica in questioni di probabilita, In the journal '. *Bollettino dei Docenti di Matematica*, *35*, 87–102.

Gal, I., & Geiger, V. (2022). Welcome to the era of vague news: A study of the demands of statistical and mathematical products in the COVID-19 pandemic media. *Educational Studies in Mathematics*, *109*(1), 1–15. DOI: 10.1007/s10649-022-10151-7 PMID: 35496813

Gallagher, T., Pfannkuch, M., & Wild, C. J. (2021). Statistical reasoning in secondary education: A case-based approach. *International Journal of Mathematics Education*, *54*(7), 879–895.

Gao, C. X., Dwyer, D., Zhu, Y., Smith, C. L., Du, L., Filia, K. M., Bayer, J., Menssink, J. M., Wang, T., Bergmeir, C., Wood, S., & Cotton, S. M. (2023). An overview of clustering methods with guidelines for application in mental health research. *Psychiatry Research*, *327*, 115265. DOI: 10.1016/j.psychres.2023.115265 PMID: 37348404

Gao, C., Zheng, Y., Wang, W., Feng, F., & He, X. (2024). Causal inference in recommender systems: A survey and future directions. *ACM Transactions on Recommender Systems*, *14*(1), 1–35. DOI: 10.1145/3568022

Garcia, J., Villavicencio, G., Altimiras, F., Crawford, B., Soto, R., Minatogawa, V., Franco, M., Martínez-Muñoz, D., & Yepes, V. (2022). Machine learning techniques applied to construction: A hybrid bibliometric analysis of advances and future directions. *Automation in Construction*, *142*, 104532. DOI: 10.1016/j.autcon.2022.104532

Garfield, J., & Ben-Zvi, D. (2008). *Developing Students' Statistical Reasoning: Connecting Research and Teaching Practice*. Springer.

Garfield, J., & Ben-Zvi, D. (2021). Statistical literacy and reasoning in the digital age. *Journal of Statistics Education : An International Journal on the Teaching and Learning of Statistics*, *29*(3), 142–156.

Ghahramani, S. (2024). *Fundamentals of Probability*. Springer.

Gilbert, J. K. (2010). *The role of visual representations in the learning and teaching of science: An introduction. Asia-Pacific Forum on Science Learning and Teaching, 11(1).* Foreword.

Goldin, G. A. (1998). Representational systems, learning, and 450 problem solving in mathematics. *The Journal of Mathematical Behavior, 451*(17), 137–165. DOI: 10.1016/S0364-0213(99)80056-1

Goldin, G. A., & Shteingold, N. (2021). Representation in mathematics education: Challenges and opportunities. *The Journal of Educational Research, 63*(5), 467–482.

Gonzalez, O., Isoda, M., & Araya, R. (2020). A new framework for statistical thinking in the time of big data and the digital economy. *Journal of Southeast Asian Education, 1*(1), 61–70.

Gopalakrishna, V. & Venugopal, K. (2023). *Assessment on the Intricacies of English Speaking Abilities by Rural Students in India.* In A. Ninawe, M. Tomar, M. S. Shende, P. Agrawal, K. M. Jaitwar, & N. S. Tiwari (Eds.), *English Language Competence in Rural Youth: Improving Job Readiness* (pp. 114-126). Current Publications.

Graham, R. L., Grotschel, M., & Lovasz, L. (1995). *Handbook Of Combinatorics.* North-Holland.

Gras, R., & Bodin, A. (2017). L'A.S.I., Analyseur et révélateur de la complexité cognitive taxonomique. 9ème Colloque International sur Analyse Statistique Implicative, Belfort – France, In Jean-Claude Régnier, Régis Gras, Raphaël Coutourier, Antoine Bodin (Eds.), pp. 128-142.

Gras, R., & Bodin, A. (2017). L'A.S.I., Analyseur et révélateur de la complexité cognitive taxonomique. 9ème Colloque International sur Analyse Statistique Implicative, Belfort – France, In Jean-Claude Régnier, Régis Gras, Raphaël Coutourier, Antoine Bodin (edus) pp. 128-142.

Gras, R., Peter, P., Briand, H., & Philippé, J. (1997). Implicative Statistical Analysis. In C. Hayashi, N. Ohsumi, N. Yajima, Y. Tanaka, H. Bock, Y. Baba (Eds.). Proceedingsofthe5th Conference of the International Federation of Classication Societies, Volume 2, pp.412-419. Tokyo, Berlin, Heidelberg, New York: Springer-Verlag.

Gras, R., Peter, P., Briand, H., & Philippé, J. (1997). Implicative Statistical Analysis. In C. Hayashi, N. Ohsumi, N. Yajima, Y. Tanaka, H. Bock, Y. Baba (Eds.). Proceedingsofthe5th Conference of the International Federation of Classification Societies, Volume 2, pp.412-419. Tokyo, Berlin, Heidelberg, New York: Springer-Verlag.

Groth, R. E. (2019). Teaching statistics in school mathematics: Challenges and opportunities. *Mathematics Education Review, 41*(1), 12–34.

Groth, R. E., Austin, J. W., Naumann, M., & Rickards, M. (2021). Toward a theoretical structure to characterize early probabilistic thinking. *Mathematics Education Research Journal, 33*(2), 1–20. DOI: 10.1007/s13394-019-00287-w

Gunderson, D., & Moore, J. O. (2008). Group learning pedagogy and group selection. *International Journal of Construction Education and Research*, *4*(1), 34–45. DOI: 10.1080/15578770801943893

Guo, J. (2020). A systematic review of visual displays in K-12 education: Effects on comprehension, retention, and higher-order learning. *Educational Psychology Review*, *32*(1), 45–67.

Gupta, S. K., & Sengupta, N. (2021). Webinar as the future educational tool in higher education of India: A survey-based study. *Technology, Knowledge and Learning*, *26*(2), 1–15. DOI: 10.1007/s10758-021-09493-7

Hair, J. F., (1995). *Multivariate Data Analysis With Reading* (4th ed.). Prentice Hall.

Hair, J., Anderson, R., Tatham, R., & Black, W. (1995). *Multivariate Data Analysis With Readings*. Prentice-Hall International, Inc.

Hair, J., Black, B., Babin, B., Anderson, E. R., & Tatham, R. L. (2005). *Multivariate Data Analysis*. Pearson Education Inc.

Halaweh, M. (2023). ChatGPT in education: Strategies for responsible implementation. *Contemporary Educational Technology*, *15*(1), 1–15. DOI: 10.30935/cedtech/13036

Harefa D H, Hulu F. (2024). Mathematics learning strategies that support Pancasila moral education: Practical approaches for teachers. Afore: Jurnal Pendidikan Matematika.

Hasim, S. M., Rosli, R., & Halim, L. (2024). A systematic review on teaching strategies for fostering students' statistical thinking. *International Journal of Learning. Teaching and Educational Research*, *23*(1), 136–158. Advance online publication. DOI: 10.26803/ijlter.23.1.8

Hausmann, R. G., & VanLehn, K. (2007). Exploring the benefits of learner-generated visual explanations for retention and application. *Learning and Instruction*, *17*(2), 172–184.

Hawes, Z. C. K., Gilligan-Lee, K. A., & Mix, K. S. (2022). Effects of spatial training on mathematics performance: A meta-analysis. *Developmental Psychology*, *58*(1), 112–137. DOI: 10.1037/dev0001281 PMID: 35073120

Henderson, P. B. (1990). *Discrete mathematics as a precursor to programming.* ACM SIGCSE Bulletin., DOI: 10.1145/323410.319070

Herreid, C. F., & Schiller, N. A. (2022). Case-based learning in STEM education: Lessons from the field. *Innovations in Education and Teaching International, 59*(1), 19–32.

Hochschild, A. R. (2003). *The second shift: Working families and the revolution at home* (Rev. ed.). Penguin Books.

Hokor, E. K., & Sedofia, J. (2021). Developing probabilistic reasoning in preservice teachers: Comparing the learner-centered and teacher-centered approaches of teaching. *Journal of Studies in Education and Science, 11*(2), 1–15. DOI: 10.19084/jsses.1114636

Hosseini, M., Rasmussen, L. M., & Resnik, D. B. (2024). Using AI to write scholarly publications. *Accountability in Research, 31*(7), 715–723. DOI: 10.1080/08989621.2023.2168535 PMID: 36697395

Hu, T., Long, C., & Xiao, C. (2021). A novel visual representation on text using diverse conditional GAN for visual recognition. *IEEE Transactions on Image Processing, 30*, 10353–10364. DOI: 10.1109/TIP.2021.3061927 PMID: 33667160

Ingram, J. (2024). Randomness and probability: Exploring student teachers' conceptions. *Mathematical Thinking and Learning, 26*(1), 1–20. DOI: 10.1080/10986065.2021.2016029

Iyer, S. (2005). Color-coded graphs for effective data interpretation in competitive exams. *Indian Educational Psychology Review, 14*(4), 90–105.

Joshi, S., & Rathod, V. (2009). Using flowcharts to improve retention and application in high school statistics. *Indian Journal of Pedagogical Research, 12*(2), 75–83.

Kaiser, H. F. (1960). The application of electronic computers to factors analysis. *Educational and Psychological Measurement, 20*(1), 141–151. DOI: 10.1177/001316446002000116

Kaiser, H. F. (1974). An index of vactorial simplicity. *Psychometrika, 39*(1), 31–36. DOI: 10.1007/BF02291575

Kamal, T., & Illiyan, A. (2021). School Teachers' Perception and Challenges Towards Online Teaching During COVID-19 Pandemic in India: An Econometric Analysis. *Asian Association of Open Universities Journal, 16*(1), 1–18. DOI: 10.1108/AAOUJ-10-2021-0122

Kapetanopoulou, P., Kouroutzi, A., & Anastasiadou, S. (2021). The Impact of Information Systems Implementation in the Greek Manufacturing Enterprises. *Applied Sciences (Basel, Switzerland)*, *11*(24), 11781. DOI: 10.3390/app112411781

Kaput, J. (1987). Representation and problem solving: methodological issues related to modelling. In Silver, E. A. (Ed.), *Teaching and Learning Mathematical Problem Solving: Multiple Research Perspectives* (pp. 381–398). Erlbaum.

Kaput, J. (1987). Representation system and mathematics. In Janvier, C. (Ed.), *Problems of representation in the teaching and learning of mathematics* (pp. 19–26). Lawrence Erlbaum Associates.

Kaur, B. (2019). The Singapore Mathematics Curriculum: Connections to TIMSS. *International Journal of Educational Research*, *95*, 180–189.

Kazak, S., & Pratt, D. (2021). Developing the role of modelling in the teaching and learning of probability. *Research in Mathematics Education*, *23*(2), 1–20. DOI: 10.1080/14794802.2020.1802328

Kiefer, M., Pielke, L., & Trumpp, N. M. (2022). Differential temporo-spatial pattern of electrical brain activity during the processing of abstract concepts related to mental states and verbal associations. *NeuroImage*, *247*, 118746. DOI: 10.1016/j.neuroimage.2022.119036 PMID: 35219860

Kilag, O. K. T., Malbas, M. H., & Miñoza, J. R. (2024). The views of the faculty on the effectiveness of teacher education programs in developing lifelong learning competence. *Higher Education Studies*, *14*(1), 1–12.

Klein, P. (2019). Visual attention on digitally represented graphs and its impact on cross-domain learning. *Journal of the Learning Sciences*, *28*(3), 200–217.

Kliegr, T., Bahník, Š., & Fürnkranz, J. (2021). A review of possible effects of cognitive biases on interpretation of rule-based machine learning models. *Artificial Intelligence*, *300*, 103553. DOI: 10.1016/j.artint.2021.103458

Koga, S. (2022). Characteristics of statistical literacy skills from the perspective of critical thinking. *Teaching Statistics*, *44*(1), 1–8. DOI: 10.1111/test.12302

Konold, C., & Lehrer, R. (2022). Rethinking data in the classroom: A theory of data modeling. *. *Educational Researcher*, *51*(2), 83–95. DOI: 10.3102/0013189X211069044

Koparan, T. (2022). The impact of a game and simulation-based probability learning environment on the achievement and attitudes of prospective teachers. *International Journal of Mathematical Education in Science and Technology*, *53*(4), 531–546. DOI: 10.1080/0020739X.2020.1868592

Korkut, E. H., & Surer, E. (2023). Visualization in virtual reality: A systematic review. *Virtual Reality (Waltham Cross)*, *27*(1), 1–20. DOI: 10.1007/s10055-023-00753-8

Kossek, E. E., & Ozeki, C. (1998). Work–family conflict, policies, and the job–life satisfaction relationship: A review and directions for organizational behavior–human resources research. *The Journal of Applied Psychology*, *83*(2), 139–149. DOI: 10.1037/0021-9010.83.2.139

Kougias, I., Seremeti, L., & Kalogeras, D. (2012). Information flow among dynamic and distributed educational environments: A formal framework. *Proceedings of the Innovative Approaches in Education – Design and Networking Conference*, 164-172.

Kougias, I., Seremeti, L., & Kalogeras, D. (2015). Occurrent vs continuant business enterprises: An ontological approach. *Proceedings of the 4th International Conference in Quantitative and Qualitative Methodologies in the Economic and Administrative Sciences*, 209-215.

Kougias, I., Seremeti, L., & Kalogeras, D. (2016). An innovative resilient strategy towards creatively teaching mathematics. *International Journal for Talent Development and Creativity*, *4*, 155–163.

Kronberger G, Burlacu B, Kommenda M, Winkler S M. (2024). "ymbolic Regression.

Kruschke, J. K. (2021). Bayesian analysis reporting guidelines. *Nature Human Behaviour*, *5*(1), 1–10. PMID: 34400814

Kumar, D., Haque, A., Mishra, K., & Islam, F. (2023). *Exploring the transformative role of artificial intelligence and metaverse in education: A comprehensive review*. Basic and Applied Sciences.

Kumar, D., Haque, A., Mishra, K., Islam, F., Kumar Mishra, B., & Ahmad, S. (2023). Exploring the transformative role of artificial intelligence and metaverse in education: A comprehensive review. *Basic and Applied Sciences*, *3*(1), 1–15. DOI: 10.56294/mr202355

Kumar, R. (2015). E-learning platforms and visual tools for statistical literacy in rural settings. *Journal of Rural Education Research*, *40*(2), 120–136.

Lapitan, L. D. S.Jr, Tiangco, C. E., Sumalinog, D A G., Sabarillo, N. S., & Diaz, J. M. (2021). An effective blended online teaching and learning strategy during the COVID-19 pandemic. *Education for Chemical Engineers*, *35*, 116–131. DOI: 10.1016/j.ece.2021.01.012

Laupichler, M. C., Aster, A., Schirch, J., & Raupach, T. (2022). Artificial Intelligence Literacy in Higher and Adult Education: A Scoping Literature Review. *Computers & Education, 187*, 1–13. DOI: 10.1016/j.compedu.2022.104441

Lavidas, K., Apostolou, Z., & Papadakis, S. (2022). Challenges and opportunities of mathematics in digital times: Preschool teachers' views. *Education Sciences, 12*(11), 1–12. DOI: 10.3390/educsci12070459

Lee, G. G., Shi, L., Latif, E., Gao, Y., & Bewersdorff, A. (2023). "Multimodality of AI for education: Towards artificial general intelligence." arXiv preprint arXiv:2301.01521.

Lerman, I. C. (1981). *Classification et Analyse Ordinale des Données*. Dunod.

Lerman, I. C. (1981). *Classification et analyse ordinale des donnes*. Dunod.

Liu, S., Liu, S., Liu, Z., Peng, X., & Yang, Z. (2022). Automated detection of emotional and cognitive engagement in MOOC discussions to predict learning achievement. *Computers & Education, 185*, 104442. DOI: 10.1016/j.compedu.2022.104461

López, J., Rodríguez, R., & Suárez, L. (2019). The impact of technology-enhanced learning in statistics education. *Computers & Education, 137*, 1–12.

Loukas, D., Papadimitriou, I., Drosos, G., & Anastasiadou, S. (2008). Investigation of a categorical data distribution via the methods of multivariate data analysis – An application to a stochastic multicriteria group decision-making problem. 67th Meeting of the European Working Group "Multiple Criteria Decision Aiding", Rovaniemi, Lapland, Finland.

Loukas, D., Papadimitriou, I., Drosos, G., & Anastasiadou, S. (2008). *Investigation of a categorical data distribution via the methods of multivariate data analysis – An application to a stochastic multicriteria group decision-making problem*. In Risto Lahdelma, Kaisa Mierrinen, Pekka Salminren, Ahti Salo (Eds). Proceedings of the 67th Meeting of the European Working Group "Multiple Criteria Decision Aiding", Rovaniemi, Lapland, Finland, pp. 151-162.

Lovasz, L. (1993). *Combinatorial Problems and Exercises* (2nd ed.). North-Holland.

Lucena, E. R F C., & Silva, C A T. (2021). The Influence of Cognitive Ability on Cognitive Biases Generated by the Representativeness Heuristic. *Revista Brasileira de Gestão de Negócios, 73*(2), 1–15. DOI: 10.7819/rbgn.v23i1.4090

Lyle, C., Rowland, M., & Ostrovski, G. (2021). *On the effect of auxiliary tasks on representation dynamics*. Machine Learning and Statistics.

Mainali, B. (2021). Representation in teaching and learning mathematics. [ERIC.]. *Journal of Education in Mathematics, Science and Technology*, *10*(2), 1–12.

Marge, M., Espy Wilson, C., Ward, N. G., Alwan, A., Artzi, Y., Bansal, M., Blankenship, G., Chai, J., Daumé, H.III, Dey, D., Harper, M., Howard, T., Kennington, C., Kruijff-Korbayová, I., Manocha, D., Matuszek, C., Mead, R., Mooney, R., Moore, R. K., & Yu, Z. (2022). Spoken language interaction with robots: Recommendations for future research. *Computer Speech & Language*, *74*, 102375. DOI: 10.1016/j.csl.2021.101255

Martín, M., & Valiña, M. D. (2023). Heuristics, biases and the psychology of reasoning: State of the art. *Psychology (Irvine, Calif.)*, *14*(2), 1–15. DOI: 10.4236/psych.2023.142016

Martynenko, G. (2010). Semiotics of statistics. *Journal of Quantitative Linguistics*, *10*(2), 105–115. DOI: 10.1076/jqul.10.2.105.16712

Mayer, R. E. (1990). The promise of multimedia learning: Using the same instructional design principles across different media. *Educational Psychologist*, *25*(4), 243–267.

Mayer, R. E. (2020). *Multimedia learning: Principles and applications*. Cambridge University Press. DOI: 10.1017/9781316941355

Mehta, A. (2012). ICT-enabled learning strategies for retention in middle schools. *International Journal of EdTech Research*, *15*(6), 145–160.

Memon M A, Ramayah T, Cheah J H. (2021). PLS-SEM statistical programs: a review. Journal of Applied Structural Equation Modeling.

Mertler C A, Vannatta R A, LaVenia K N. (2021). Advanced and multivariate statistical methods: Practical application and interpretation.

Michel Villarreal, R., & Vilalta-Perdomo, E. (2023). Challenges and opportunities of generative AI for higher education as explained by ChatGPT. *Education*, *11*(2), 1–15. DOI: 10.3390/educ11020001

Ministry of Education. (2020). National Education Policy 2020. *Government of India*. https://www.education.gov.in/sites/upload_files/mhrd/files/NEP_Final_English_0.pdf

Mittal, U., Sai, S., Chamola, V., & Sangwan, D. (2024). A comprehensive review on generative AI for education. *IEEE Access : Practical Innovations, Open Solutions*, *12*, 142733–142759. DOI: 10.1109/ACCESS.2024.3468368

Mogavi, R. H., Deng, C., Kim, J. J., Zhou, P., & Kwon, Y. D. (2023). "Exploring user perspectives on ChatGPT: Applications, perceptions, and implications for AI-integrated education." arXiv preprint arXiv:2301.08221.

Morales-Vargas, S. R., & Benitez-Medina, R. (2024). The use of artificial intelligence as a learning tool. A first focus on solving statistics problems. *Proceedings of the 17th Annual International Conference of Education Research and Innovation*, 3412-3416, Seville, Spain. DOI: 10.21125/iceri.2024.0882

Mutohhari, F., Sutiman, S., Nurtanto, M., & Kholifah, N. (2021). Difficulties in Implementing 21st Century Skills Competence in Vocational Education Learning. *Journal of Research in Education*, *31*(2), 1–13.

National Governors Association Center for Best Practices & Council of Chief State School Officers. (2010). *Common Core State Standards for Mathematics.*

Nelson, B. (2021). *Foundations and methods of stochastic simulation.* Springer. DOI: 10.1007/978-3-030-86194-0

Nicolaou, S., Gagatsis, A., Deliyianni, E., Panaoura, A., Elia, I., & Anastasiadou, S. (2017). Tracing the beliefs and self-efficacy beliefs of undergraduate economic sciences students: the case of the representations of functions. *Proceedings* of 9ème Colloque International sur Analyse Statistique Implicative. Belfort – France, pp. 461-478.

Nikou, S., & Aavakare, M. (2021). An Assessment of the Interplay Between Literacy and Digital Technology in Higher Education. *Education and Information Technologies*, *26*(6), 6715–6733. DOI: 10.1007/s10639-021-10451-0

Northrup, J. M., Vander Wal, E., Bonar, M., Fieberg, J., Laforge, M. P., Leclerc, M., Prokopenko, C. M., & Gerber, B. D. (2022). Conceptual and methodological advances in habitat-selection modeling: Guidelines for ecology and evolution. *Ecological Applications*, *32*(1), 1–15. DOI: 10.1002/eap.2470 PMID: 34626518

Noyes, A., Wake, G., & Drake, P. (2014). *Bowland Maths Evaluation*. University of Nottingham Centre for Research in Mathematics Education.

Ntotsi, P., & Anastasiadou, S. D. (2019a). Contemporary advanced statistical methods for the science of educational research: Principal components analysis versus L' analysee factorielle des correspondances. *New Trends and Issues Proceedings on Humanities and Social Sciences*. [Online]. 6(1), pp 246–255. Available from from: www.prosoc.eu

Ntotsi, P., & Anastasiadou, S. D. (2019[a]). Contemporary advanced statistical methods for the science of educational research: Principal components analysis versus L' analysee factorielle des correspondances. *New Trends and Issues Proceedings on Humanities and Social Sciences*. [Online]. 6(1), pp 246–255. Available from from: www.prosoc.eu

Ntotsi, P., & Anastasiadou, S. D. (2019b). Comparison of multivariate pattering methods in group/cluster identification regarding the science of educational research: Implicative Statistical Analysis vs. L' Analysee Factorielle des Correspondances. *New Trends and Issues Proceedings on Humanities and Social Sciences.* [Online]. 6(1), pp 238–245. Available from from: www.prosoc.eu

Ntotsi, P., & Anastasiadou, S. D. (2019b). Comparison of multivariate pattering methods in group/cluster identification regarding the science of educational research: Implicative Statistical Analysis vs. L' Analysee Factorielle des Correspondances. *New Trends and Issues Proceedings on Humanities and Social Sciences.* [Online]. 6(1), pp 238–245. www.prosoc.eu

Nusantara, T., Rahmatina, D., & Purnomo, H. (2021). The Statistical Creative Framework in Descriptive Statistics Activities. *International Journal of Instruction, 14*(3), 1–15.

O'Regan, G., Gries, D., & Schneider, F. B. (2016). *Guide to Discrete Mathematics.* Springer. DOI: 10.1007/978-3-319-44561-8

OECD. (2019). *PISA 2018 Results (Volume I): What Students Know and Can Do.* OECD Publishing.

Otsuka, J. (2023). *Thinking about Statistics: The Philosophical Foundations.* Routledge., DOI: 10.4324/9781003319061

Page, R. L. (2003). *Software is discrete mathematics.* ACM SIGPLAN International. DOI: 10.1145/944705.944713

Pang, Y., Xiao, F., Wang, H., & Xue, X. (2014). A clustering-based grouping model for enhancing collaborative learning. *Proceedings of the 13th International Conference on Machine Learning and Applications*, Detroit, MI, USA, 562-567. DOI: 10.1109/ICMLA.2014.94

Panitsides, E., & Anastasiadou, S. (2015). Lifelong Learning Policy Agenda in the European Union: A bi- level analysis. *Open Review of Educational Research, 2*(1), 128–142. DOI: 10.1080/23265507.2015.1043936

Papadaki, Z. E., & Anastasiadou, S. D. (2019). Evaluating perception, expectation of consumers, and service quality gap in Greek banking in a period of financial crisis and capital controls. In N. Sykianakis, P. Polychronidou, & A. Karasavvoglou (Eds.), Economic and Financial Challenges for Eastern Europe (pp. 67–80). Springer Pro-ceedings in Business and Economics. Springer, Cham. https://doi.org/DOI: 10.1007/978-3-030-12169-3_5

Papadaki, Z., Anastasiadou, S., Masouras, A., & Papalexandris, S. (2023). Factors Influencing the Use of Digital Marketing by Telemedicine Services. 10th International Conference on Strategic Innovative Marketing and Tourism. Zakynthos, Ionian Islands, Greece. In A. Kavoura et al. (eds.), Strategic Innovative Marketing and Tourism, Springer Proceedings in Business and Economics, (2023), 855-862. DOI: 10.1007/978-3-031-51038-0_92

Peeck, J. (1987). The role of delayed recall in enhancing retention through visual instruction. *Journal of Educational Psychology*, *79*(2), 128–133.

Petridou, A., Elia, I., Gagatsis, A., & Anastasiadou, S. (2017). L'appréhension perceptive des figures géométrique. *Proceedings* of 9ème Colloque International sur Analyse Statistique Implicative. Belfort – France, pp. 231-247.

Pfannkuch, M., & Wild, C. J. (2020). Statistical thinking in education: Challenges and opportunities. *Statistics Education Research Journal*, *19*(2), 45–62.

Pitt, B., Gibson, E., & Piantadosi, S. T. (2022). Exact number concepts are limited to the verbal count range. *Psychological Science*, *33*(1), 1–13. DOI: 10.1177/09567976211034502 PMID: 35132893

Planinic, M. (2013). Graphical representations as a bridge between theoretical knowledge and real-world problem-solving in higher education. *Physical Review Special Topics. Physics Education Research*, *9*(1), 1–14.

Pospiech, G., Merzel, A., Zuccarini, G., & Weissman, E. (2021). "The role of mathematics in teaching quantum physics at high school." Physics: From Research to Teaching, Springer, 1-15. DOI: 10.1007/978-3-030-78720-2_4

Poultsakis, S., Papadakis, S., Kalogiannakis, M., & Psycharis, S. (2021). The management of digital learning objects of natural sciences and digital experiment simulation tools by teachers. *Advances in Mobile and Distance Learning*, *1*(2), 58–71. DOI: 10.25082/AMLER.2021.02.002

Pranaya, D., & Venugopal, K. (2023) Impact of Mother Tongue on Students' speaking Abilities in English Language at Rural Places. In A. Ninawe, M. Tomar, M. S. Shende, P. Agrawal, K. M. Jaitwar, & N. S. Tiwari (Eds.), *English Language Competence in Rural Youth: Improving Job Readiness* (pp. 143-157). Current Publications.

Pratiwi, D. I., & Waluyo, B. (2023). Autonomous Learning and the Use of Digital Technologies in Online English Classrooms in Higher Education. *Contemporary Educational Technology, 15*(2), ep423. DOI: 10.30935/cedtech/13094

Quinones, D., Ruz, F., Diaz-Arancibia, J., Paz, F., Osega, J., & Rojas, L. F. (2024). Innovating statistics education: The design of a novel app using design thinking. *Applied Sciences (Basel, Switzerland), 14*(18), 8515. DOI: 10.3390/app14188515

Rahate, A., Walambe, R., Ramanna, S., & Kotecha, K. (2022). Multimodal co-learning: Challenges, applications with datasets, recent advances and future directions. *Information Fusion, 81*, 203–239. DOI: 10.1016/j.inffus.2021.12.003

Rahman, M. M. (2023). Sample size determination for survey research and non-probability sampling techniques: A review and set of recommendations. *Journal of Entrepreneurship, Business and Management, 11*(1), 1–18. DOI: 10.5281/zenodo.7346384

Ramezanali, N. (2014). *Multimodal instructional strategies and elaborative interrogation methods in education*. ERIC Database.

Rane, N. L., Choudhary, S. P., & Rane, S. (2024). Education 4.0 and 5.0: Integrating artificial intelligence (AI) for personalized and adaptive learning. *Journal of Artificial Intelligence and Robotics, 1*(1), 29–43. DOI: 10.61577/jaiar.2024.100006

Rasul, T., Nair, S., Kalendra, D., & Robin, M. (2023). The role of ChatGPT in higher education: Benefits, challenges, and future research directions. *Journal of Applied Learning and Teaching, 6*(1), 1–12. DOI: 10.37074/jalt.6.1.10

Reddy, C.K. and Vinzamuri, B. (2018). A survey of partitional and hierarchical clustering algorithms. *Data Clustering*, 87-110.

Reddy, D., Pulluru, K., & Sharma, R. (2024). *Principles of Statistics & Research Methodology*. AGPHBooks. ISBN: 978-81-19843-21-3.

Relkin, E., de Ruiter, L. E., & Bers, M. U. (2021). Learning to code and the acquisition of computational thinking by young children. *Computers & Education, 172*, 104224. DOI: 10.1016/j.compedu.2021.104222

Ridgway, J. (2016). Implications of the data revolution for statistics education. *International Statistical Review, 84*(3), 528–549. DOI: 10.1111/insr.12110

Ridgway, J. (2019). Statistical literacy and big data: Implications for education. *Statistics Education Research Journal, 18*(1), 29–46.

Riffenburgh, R. H. (2012). *Statistics in Medicine*. Elsevier.

Ringle, C. M., Sarstedt, M., Sinkovics, N., & Sinkovics, R. R. (2023). A perspective on using partial least squares structural equation modelling in data articles. *Data in Brief, 48*, 109074. DOI: 10.1016/j.dib.2023.109074 PMID: 37066088

Roediger, H. L. (2006). Visual testing approaches for enhancing retention and material integration. *Psychological Science, 17*(12), 1063–1070.

Rosen, K. H. (2019). *Discrete Mathematics and Its Applications* (8th ed.). McGraw-Hill.

Roy, M., & Patel, A. (2013). Alleviating math anxiety with animated visuals during statistical problem-solving. *Indian Journal of Cognitive Psychology, 22*(3), 56–70.

Rubin, A., & Justice, L. (2019). Language barriers in statistical education: Supporting multilingual learners. *Statistics Education Research Journal, 18*(1), 22–38.

Ruiz-Rojas, L. I., Acosta-Vargas, P., De-Moreta-Llovet, J., & Gonzalez-Rodriguez, M. (2023). Empowering education with generative artificial intelligence tools: Approach with an instructional design matrix. *Sustainability (Basel), 15*(15), 11524. DOI: 10.3390/su151511524

Sailer, M., Schultz-Pernice, F., & Fischer, F. (2021). Contextual facilitators for learning activities involving technology in higher education: The C-model. *Computers in Human Behavior, 121*, 103023. DOI: 10.1016/j.chb.2021.106794

Santos, A. R., & Gross, M. U. (2021). Equity in STEM education: Addressing intersectionality through culturally responsive pedagogy. *Journal of Educational Enquiry, 18*(2), 67–89.

Saputra, I., Astuti, M., Sayuti, M., & Kusumastuti, D. (2023). Integration of artificial intelligence in education: Opportunities, challenges, threats and obstacles. A literature review. *The Indonesian Journal of Computer Science, 12*(4), 1590–1600. DOI: 10.33022/ijcs.v12i4.3266

Schene, A., Wijngaarden, B., & Koeter, M. (1998). Family Care giving in Schizophrenia: Domains, Distress. *Schizophrenia Bulletin, 24*(4), 609–618. DOI: 10.1093/oxfordjournals.schbul.a033352 PMID: 9853792

Schölkopf, B., Locatello, F., Bauer, S., & Ke, N. R. (2021). Toward causal representation learning. *Proceedings of the IEEE*.

Seremeti, L., Anastasiadis, L., & Anastasiadou, S. (2024). The role of data in Environmental, Social and Governance strategy. In *Papademetriou, C., Anastasiadou, S. Ragazou, K., Garefalakis, A., and Papalexandris, S. (2024)* (pp. 201–218). ESG and Total Quality Management in Human Resources. IGI Global Publisher of Timely Knowledge., DOI: 10.4018/979-8-3693-2991-7.ch010

Seremeti, L., & Kougias, I. (2013). Networks of aligned ontologies through a category theoretic approach. *Journal of Emerging Trends in Computing and Information Sciences*, *4*(10), 785–794.

Shaik, T., Tao, X., Li, Y., Dann, C., McDonald, J., Redmond, P., & Galligan, L. (2022). A review of the trends and challenges in adopting natural language processing methods for education feedback analysis. *IEEE Access : Practical Innovations, Open Solutions*, *10*, 56720–56739. DOI: 10.1109/ACCESS.2022.3177752

Sharma, H. (2021). Statistical significance or clinical significance? A researcher's dilemma for appropriate interpretation of research results. *Saudi Journal of Anaesthesia*, *15*(4), 431–434. DOI: 10.4103/sja.sja_158_21 PMID: 34658732

Sharma, N. (2003). Engagement and retention through culturally relevant visual examples among tribal learners. *Journal of Indigenous Education Studies*, *10*(3), 20–35.

Sharma, P. (2016). Collaborative graph-based learning in urban schools. *Educational Technology Insights*, *28*(1), 50–63.

Sharma, S. (1996). *Applied Multivariate Techniques*. John Willey & Sons, Inc.

Shodiqin, A., Sukestiyarno, Y. L., Wardono, W., & Isnarto, I. (2021). Probabilistic Thinking Profile of Mathematics Teacher Candidates in Problem Solving Based on Self-Regulated Learning. *European Journal of Educational Research*, *10*(2), 1–15. DOI: 10.12973/eu-jer.10.3.1199

Sikstrom, S., & Garcia, D. (2020). *Statistical Semantics: Methods and Applications*. Springer., DOI: 10.1007/978-3-030-37250-7

Singh, A. (2018). Interactive diagrams and their impact on statistical application skills in Indian undergraduate programs. *Indian Journal of Education Studies*, *35*(4), 89–102.

Sinha, R., & Kapoor, D. (2010). Alignment of textbook visuals with statistical learning outcomes in Indian curricula. *Journal of Curriculum Studies*, *18*(5), 89–102.

Sitopu J W, Khairani M, Roza M, Judijanto L. (2024). The Importance of Integrating Mathematical Literacy in the Primary Education Curriculum: A Literature Review. International Journal of Teaching and Learning, 1-18.

Skivington, K., Matthews, L., Simpson, S. A., Craig, P., Baird, J., Blazeby, J. M., Boyd, K. A., Craig, N., French, D. P., McIntosh, E., Petticrew, M., Rycroft-Malone, J., White, M., & Moore, L. (2021). Framework for the development and evaluation of complex interventions: Gap analysis, workshop and consultation-informed update. *Health Technology Assessment*, *25*(27), 1–128. DOI: 10.3310/hta25570 PMID: 34590577

Sohail, S. S., Farhat, F., Himeur, Y., Nadeem, M., Madsen, D. Ø., Singh, Y., Atalla, S., & Mansoor, W. (2023). Decoding ChatGPT: A taxonomy of existing research, current challenges, and possible future directions. *Journal of King Saud University. Computer and Information Sciences*, *35*(8), 101675. DOI: 10.1016/j.jksuci.2023.101675

Souravlas, S. Anastasiadou, S. & Kostoglou, I. (2022b). A Novel Method For General Hierarchical System Modeling Via Colored Petri Nets Based on Transition Extractions From Real Data Sets. Appl. Sci. 2023, 13, 339. DOI: 10.3390/app13010339

Souravlas, S., & Anastasiadou, S. (2020). Pipelined Dynamic Scheduling of Big Data Streams. *Applied Sciences (Basel, Switzerland)*, *10*(14), 4796. DOI: 10.3390/app10144796

Souravlas, S., & Anastasiadou, S. (2022). On Implementing Social Community Clouds Based on Markov Models. *IEEE Transactions on Computational Social Systems*. Advance online publication. DOI: 10.1109/TCSS.2022.3213273

Souravlas, S., Anastasiadou, S. D., Tantalaki, N., & Katsavounis, S. (2022). A Fair, Dynamic Load Balanced Task Distribution Strategy for Heterogeneous Cloud Platforms Based on Markov Process Modeling. *IEEE Access : Practical Innovations, Open Solutions*, *10*, 26149–26162. DOI: 10.1109/ACCESS.2022.3157435

Souravlas, S., Anastasiadou, S., Economides, Th., & Katsavounis, S. (2023). Probabilistic Community Detection in Social Networks. *IEEE Access : Practical Innovations, Open Solutions*, *11*, 25629–25641. DOI: 10.1109/ACCESS.2023.3257021

Souravlas, S., Anastasiadou, S., & Katsavounis, S. (2021). More on pipelined dynamic scheduling of big data streams. *Applied Sciences (Basel, Switzerland)*, *11*(1), 61. DOI: 10.3390/app11010061

Souravlas, S., Anastasiadou, S., & Katsavounis, S. (2021b). A Survey on the Recent Advances of Deep Community Detection. *Applied Sciences (Basel, Switzerland), 11*(16), 7179. DOI: 10.3390/app11167179

Souravlas, S., Anastasiadou, S., Tantalaki, N., & Katsavounis, S. (2021a). A dynamic load balanced task distribution strategy for heterogeneous cloud platforms based on Markov process modeling. *IEEE Access : Practical Innovations, Open Solutions, 10*, 2169–3536. https://ieeexplore.ieee.org/abstract/document/9729723

Souravlas, S., Anastasiadou, S., Tantalaki, N., & Katsavounis, S. (2021a). A Dynamic Load Balanced Task Distribution Strategy for Heterogeneous Cloud Platforms Based on Markov Process Modeling. *IEEE Access : Practical Innovations, Open Solutions, 10*, 2169–3536. https://ieeexplore.ieee.org/abstract/document/9729723

Souravlas, S., Katsavounis, S., & Anastasiadou, S. (2020). On Modeling and Simulation of Resource Allocation Policies in Cloud Computing Using Colored Petri Nets. *Applied Sciences (Basel, Switzerland), 10*(16), 5644. DOI: 10.3390/app10165644

Šrol, J. (2022). Individual differences in epistemically suspect beliefs: The role of analytic thinking and susceptibility to cognitive biases. *Thinking & Reasoning, 28*(1), 1–25. DOI: 10.1080/13546783.2021.1938220

Steed, C. A. (2025). *Interactive data visualization*. Data Analytics for Intelligent Transportation Systems.

Susilawati, E., Khaira, I., & Pratama, I. (2021). Antecedents to Student Loyalty in Indonesian Higher Education Institutions: The Mediating Role of Technology Innovation. *Kuram ve Uygulamada Egitim Bilimleri, 21*(4), 1329–1344.

Sweller, J., Ayres, P., & Kalyuga, S. (2019). *Cognitive load theory: Applications in education*. Springer.

Taraza, E., Anastasiadou, S., & Masouras, A. (2023a). Malcolm Baldrige National Quality Award (MBNQA) Quality Tool in Education: A Systematic Literature Review. Proceedings of the European Conference on Innovation and Entrepreneurship, ECIE, 2023, 1, pp. 29–38.

Taraza, E., & Anastasiadou, S. (2019). EFQM Excellence Model in Vocational Lyceum: Reliability and validity of the EFQM instrument. In *Proceedings of the 13th Annual International Technology, Education and Development Conference (INTED2019)(pp.* 2273–2285*)*. Valencia, Spain. DOI: 10.21125/inted.2019.0637

Taraza, E., Anastasiadou, S., Masouras, A., & Papademetriou, C. (2023b). Sustainable Development and Implementation of Quality Management Excellence Models in Public Organizations: A Systematic Literature Review. *Sustainability (Basel), 2023*(15), 7971. DOI: 10.3390/su15107971

Taraza, E., Anastasiadou, S., Papademetriou, C., & Masouras, A. (2024). Evaluation of Quality and Equality in Education Using the European Foundation for Quality Management Excellence Model—A Literature Review. *Sustainability (Basel), 2024*(16), 960. DOI: 10.3390/su16030960

Thelen, A., Zhang, X., Fink, O., Lu, Y., Ghosh, S., Youn, B. D., Todd, M. D., Mahadevan, S., Hu, C., & Hu, Z. (2022). A comprehensive review of digital twin—part 1: Modeling and twinning enabling technologies. *Structural and Multidisciplinary Optimization, 65*(4), 1–20. DOI: 10.1007/s00158-022-03425-4

Tinker, R., & Konold, C. (2015). *CODAP: Data Exploration Tools for the Classroom.* Concord Consortium Reports.

Tishkovskaya, S., & Lancaster, G. A. (2023). Statistical education: Current challenges and future directions. *. *Statistics Education Research Journal, 22*(1), 12–30. DOI: 10.52041/serj.v22i1.123

Tong, D. H., Uyen, B. P., & Quoc, N. V. A. (2021). The improvement of 10th students' mathematical communication skills through learning ellipse topics. *Heliyon, 7*(9), e07967. DOI: 10.1016/j.heliyon.2021.e08282 PMID: 34765795

Tufte, E. R. (2001). *The Visual Display of Quantitative Information.* Graphics Press.

Ünal, Z. E., Ala, A. M., Kartal, G., Özel, S., & Geary, D. C. (2023). Visual and symbolic representations as components of algebraic reasoning. *Journal of Numerical Cognition*, 1–18.

Ungureanu, L. (2024). Statistics in social sciences. *Journal of Research in Educational Sciences, 17*(1), 57–71. DOI: 10.14505/jres.v15.1(17).06

Valkanos, E., Giossi, S., & Anastasiadou, S. (2010). Certified Adult Educators´ Attitudes towards Continuing Education. Review of European Studies. 2(1), (2010) 61-68. Part of ISSN: 19187173 19187181.

Valsamidis, S., Mandilas, A., Florou, G., & Anastasiadou, S. (2024). The use of e-government services provided by the local authorities: The case of the municipality of Kavala. *International Journal of Electronic Governance, 16*(2), 196–218. DOI: 10.1504/IJEG.2024.140790

Van de Schoot, R., Depaoli, S., King, R., Kramer, B., Märtens, K., Tadesse, M. G., Vannucci, M., Gelman, A., Veen, D., Willemsen, J., & Yau, C. (2021). Bayesian statistics and modelling. *Nature Reviews. Methods Primers*, *1*(1), 1–15. DOI: 10.1038/s43586-020-00001-2

Van Dijke-Droogers, M., Drijvers, P., & Bakker, A. (2022). Introducing statistical inference: Design of a theoretically and empirically based learning trajectory. *International Journal of Science and Mathematics Education*, *20*(8), 1–18. DOI: 10.1007/s10763-021-10208-8

Venugopal, K. (May, 2024). Exploring Gen Z Perspectives toward ChatGPT in Academic Contexts. In R. Bansal, A. Chakir, A. HafazNgah, F. Rabby, & A. Jain (Eds.), AI Algorithms and ChatGPT for Student Engagement in Online Learning (pp. 51-69). IGI Global. https://doi.org/DOI: 10.4018/979-8-3693-4268-8.ch004

Venugopal, K., & Somanadh, K. V. (August 2024). Gen Z's Journey through Virtual Library Landscapes: Predictive Analysis. In *Libraries as Wisdom Hubs: Interdisciplinary Insights for the Future.* W.S. College of Commerce, Wardha, Maharashtra.

Venugopal, K., Das, S., & Badawy, Hassan (2023). Prediction Analysis of Gen Zers' Attitudes on Ecological Consciousness. In B. K. Mishra (Ed.), *Sustainable Science and Intelligent Technologies for Societal Development* (pp. 342-357). IGI Global.

Verma, S. (2017). Bridging theoretical and applied statistics using visual aids in competitive exam preparation. *Journal of Educational Research in India*, *12*(3), 45–55.

Watson, J. (2021). Developing statistical thinking in secondary education. *Journal of Mathematics Education*, *52*(5), 672–687.

Watson, J., & Donne, J. (2021). The role of context in students' statistical thinking. *Educational Research Review*, *33*(4), 215–230.

Weiskopf, D. (2022). Uncertainty visualization: Concepts, methods, and applications in biological data visualization. *Frontiers in Bioinformatics*, *2*, 934741. DOI: 10.3389/fbinf.2022.793819 PMID: 36304261

Wild, C. J., & Pfannkuch, M. (1999). Statistical thinking in empirical enquiry. *International Statistical Review*, *67*(3), 223–248. DOI: 10.1111/j.1751-5823.1999.tb00442.x

Xu, H., Berres, A., Liu, Y., Allen-Dumas, M. R., & Sanyal, J. (2022). An Overview of Visualization and Visual Analytics Applications in Water Resources Management. *Environmental Modelling & Software*, *147*, 105261. DOI: 10.1016/j.envsoft.2022.105396

Yau N. (2024). Visualize this: the FlowingData guide to design, visualization, and statistics.

Yau, N. (2024). *Visualize This: The FlowingData Guide to Design*. Visualization, and Statistics. DOI: 10.1002/9781394319817

Zhai, X., Chu, X., Chai, C. S., Jong, M. S. Y., Istemic, A., Spector, M., Liu, J. B., & Li, J. Y. Y. (2021). A review of artificial intelligence (AI) in education from 2010 to 2020. *Complexity*, *1*(1), 8812542. DOI: 10.1155/2021/8812542

Zheng, J., Harris, M. R., Masci, A. M., Lin, Y., Hero, A., Smith, B., & He, Y. (2016). The ontology of biological and clinical statistics (OBCS) for standardized and reproducible statistical analysis. *Journal of Biomedical Semantics*, *7*(53), 53. Advance online publication. DOI: 10.1186/s13326-016-0100-2 PMID: 27627881

Zheng, Z., Zhou, B., Zhou, D., Khan, A. Q., Soylu, A., & Kharlamov, E. (2022). Towards a statistic ontology for data analysis in smart manufacturing. *Proceedings of the 21st International Semantic Web Conference*, Hangzhou, China.

Zhou, S., Xu, H., Zheng, Z., Chen, J., Li, Z., Bu, J., & Wu, J. (2024). A comprehensive survey on deep clustering: Taxonomy, challenges, and future directions. *ACM Computing Surveys*, *56*(2), 1–37.

Ziatdinov, R., & Valles, J. R.Jr. (2022). Synthesis of modeling, visualization, and programming in GeoGebra as an effective approach for teaching and learning STEM topics. *Mathematics*, *10*(3), 398. DOI: 10.3390/math10030398

Zou, C., Li, P., & Jin, L. (2021). Online College English Education in Wuhan Against the COVID-19 Pandemic: Student and Teacher Readiness, Challenges and Implications. *PLoS One*, *16*(3), e0248305. DOI: 10.1371/journal.pone.0258137 PMID: 34597337

About the Contributors

Sofia Anastasiadou is a Professor of Statistics and Research Methodology, Biostatistics Biomedical Research Methodology and at the School Health Sciences of the University of Western Macedonia (UOWM). She is the Dean of the Health Sciences of UOWM, Senate Member, and Member of post graduate studies committee. She is also a Member of the Senate and a MODIP Member of the same University. Her teaching focuses in the field of cognitive Research Methodology (Quantitative and Qualitative), Educational Research and Statistical Data Analysis, Big Data. She has written three books on Statistics and Research Methodology and has published over 300 articles in International and Greek Journals as well as in International and Greek Conferences. She has participated in scientific dialogues/ round tables in more than 200 lectures at International and Greek Conferences. She has organized and co-organized Greek and International Scientific Conferences. She is also a Member of a number of Scientific Committees of International Conferences and is an Associate Editor in a number of International Journals. Since the spring semester 2021 she works as a Collaborative Academic Staff teaching in the Distance Learning Program of Special Education at the School of Education and Social Sciences at Frederick University.

Lamprini Seremeti is a Collaborative Academic Staff at Frederick University since 2017. She is also a Lecturer in Programming in the Department of Agricultural Economics & Rural Development at the School of Applied Economics and Social Science of the Agricultural University of Athens. Her teaching experience in Higher Education Institutions, since 2005, involves cognitive subjects such as expert systems, artificial intelligence, semantic web, business management, programming languages, calculus, algebra, statistics, differential equations, research methodology, informatics applied in pedagogy. She has also served as a teacher of Informatics, Mathematics and Special Education in Primary and Secondary Schools, since 2001.Dr Seremeti holds a PhD in Computer Science with a focus on propagating knowledge in networks of aligned ontologies from Hellenic Open University; a Master in Special Education from University of Rome; a Master in Pure Mathematics from University of Patras; a Bachelor in Mathematics from University of Patras; a Bachelor in Informatics from University of Western Greece; and a Bachelor in Law from University of Nicosia. She has also participated as a researcher in EU-funded projects, in which, her tasks involved mathematical modeling, data processing, knowledge representation and management, and conceptualization of heterogeneous environments (socio-economic, educational, biomedical, ambient intelligent). Dr Seremeti has published her research work in peer-reviewed international journals (14 articles) and book chapters (5 articles), and has presented her work in international peer-reviewed conferences (30 articles). Her research interests focus on theories applied in inclusive education, interpretation of Law in ambient intelligent environments, knowledge representation and management through ontologies, mathematical modeling.

***

Nilesh Anute is an Associate Professor at Sri Balaji University, Pune. With a strong academic background and extensive teaching experience, Dr. Anute specializes in areas such as management education, research methodologies, and strategic studies. Known for his student-centric approach and commitment to academic excellence, he actively contributes to curriculum development and academic research. Dr. Anute continues to inspire future professionals through his dynamic teaching and scholarly contributions.

Susanta Das is a seasoned educator with a proven record of teaching and mentoring diverse students, working at multiple institutions, and providing service to the various initiatives of organizations. He received his Ph.D. and M.A. degrees from Western Michigan University (WMU), USA, and M.Sc. degree from Banaras Hindu University (BHU), India all in Physics. He continued his research as a Marie-Curie post-doctoral fellow at Stockholm University, Stockholm, Sweden on beam diagnostics for the DESIREE (Double ElectroStatic Ion Ring ExpEriment) in the project DITANET (Diagnostic Techniques for particle Accelerators – a Marie-Curie initial training NETwork), at the Indian Institute of Science Education and Research-Kolkata, India on high-pressure physics as a Project Scientist-B, and at the University of Electro-Communications, Tokyo, Japan on ion-surface interactions as a Post-doctoral fellow. Throughout his career, Dr. Das worked and collaborated with many researchers, Ph.D., master, and visiting students from different countries. He further visited many countries to discuss research and an international conference participant (UK, Italy, Germany, Greece, Belgium, Bulgaria, Romania, Brazil etc.).

He co-authored several research articles in WoS/Scopus indexed international journals, conference proceedings, and scholarly book chapters published by renowned international publishing houses. Beyond his research endeavors, Dr. Das has a storied history of service to academic and administrative committees, exhibiting his commitment to the growth and development of educational institutions. His experience includes tenure at Central University South Bihar, Sri Sri University, and P.K. University before assuming his current role at Ajeenkya DY Patil University. He received the Marie-Curie post-doctoral fellowship in Sweden, Science Academies' Summer Research Fellowship in India, Gwen Frostic Doctoral Fellowship, Department Graduate Research and Creative Scholar Award by WMU, and Leo R. Parpart Doctoral Fellowship by Dept. of Physics, WMU, among many others, throughout his academic journey. At present, Dr. Das continues to delve into cutting-edge fields, with a keen interest in nanotechnology, UN SDGs, and data science. His multifaceted contributions, spanning teaching, research, and administrative leadership, showcase a dedicated professional who is instrumental in advancing the vision and mission of the institutions he serves.

Vishal Jain is presently working as Professor at Department of Computer Science and Engineering, School of Engineering and Technology, Sharda University, Greater Noida, U. P. India. Before that, he has worked for several years as an Associate Professor at Bharati Vidyapeeth's Institute of Computer Applications and Management (BVICAM), New Delhi. He has more than 15 years of experience in the academics. He obtained Ph.D (CSE), M.Tech (CSE), MBA (HR), MCA, MCP and CCNA. He has authored more than 100 research papers in reputed conferences and journals, including Web of Science and Scopus. He has authored and edited more than 45 books with various reputed publishers, including Elsevier, Springer, IET, Apple Academic Press, CRC, Taylor and Francis Group, Scrivener, Wiley, Emerald, NOVA Science, IGI-Global and River Publishers. His research areas include information retrieval, semantic web, ontology engineering, data mining, ad hoc networks, and sensor networks. He received a Young Active Member Award for the year 2012–13 from the Computer Society of India, Best Faculty Award for the year 2017 and Best Researcher Award for the year 2019 from BVICAM, New Delhi.

T.C. Manjunath was born on 6th February 1967. He did his BE from RV College of Engg. in 1989, ME in Robotics & Automation from LD College of Engg. in 1992 with First Rank and PhD in Systems & Control Engg. from IIT Bombay in the year 2007. He is currently working as the Dean Research R D and as Professor in the Dept. of Computer Science & Engineering with specialization in IoT, Cyber Security & Blockchain Technology in Rajarajeswari College of Engineering, Bengaluru, Karnataka. He has completed B.E. Degree from R.V. College of Engg.,

M.E. Govt.'s LD College of Engg., Gujarat Univ., Ahmedabad in First Rank and Ph.D. in Systems and Control Engineering from the Indian Institute of Technology Bombay IIT Bombay He has got a teaching academic, research and administrative experience of more than 35+ years 10 years as Principal, 8 years as HoD rest as faculty and has published more than 1200+ research papers in various national & international journals and conferences, with 175+ patents, 25 book-chapters, 25 awards and also published 25 text books in various field of science, engineering and technology, completed projects worth Rs. 9 Crores till date apart from guiding 10 PhD scholars. He is a member of 30 Professional Society Chapters and is a fellow of IE, IETE, AMIEE and a chartered engineer. He got the best research scholar award in 2006 in IIT Bombay amongst all the research scholars of the institute. He has got 10 awards in academics and research activities.

Harleen Kaur Dhillon is a distinguished Professor of Management at CGC Mohali. With a robust academic and corporate background spanning 17 years, she brings a wealth of experience and expertise to the institution. Her journey is marked by a blend of academic achievements, research contributions, and leadership roles that underscore her commitment to excellence in education. Her contributions to the field of education have been widely recognized. In 2024, she was honored with the Indian Glory Award for her significant contributions to student development. Additionally, she has chaired technical sessions at national conferences and seminars. Her academic qualifications include a Ph.D. from Punjab Technical University, a Master's in Business Administration, a Master's in Commerce, and a Bachelor's in Education.

Archan Mitra is an Assistant Professor at School of Media Studies (SOMS) at Presidency University, Bangalore. He is the author of two book "Cases for Classroom Media and Entertainment Business" and "Multiverse and Media", he also has other several edited books to his credit. He has done his doctorate from Visva-Bharati Santiniketan, West Bengal in the field of "environmental informatics and communication for sustainability". In addition to that he is a certified Science Communicator and Journalism from Indian Science Communication Society (ISCOS), certified Corporate Trainer with Amity Institute of Training and Development, Certified Social Media Network Analyst. He has a strong interest in environmental communication. He was awarded certificate of merit by PRSI, Kolkata Chapter and Medal of Honor by Journalistic Club of Kolkata. He was working as a research assistant with the World Bank's "Environmental Capacity Building in Southeast Asia" project at IIM Kashipur. He was instrumental in launching the World Bank's Green MBA MOOC, he has also assisted in the research project on

Uttarakhand disaster mitigation by ICSSR, the leading research on Uttarakhand disaster.

P. Selvakumar is currently an Associate Professor in the Department of Science and Humanities at Nehru Institute of Technology, Coimbatore, Tamil Nadu, India. Since 2023, he has served as the Department IQAC Coordinator at the institute. Prior to joining Nehru Institute of Technology, he held positions as an Assistant Professor in the Department of Science and Humanities various engineering colleges in Coimbatore, Tamil Nadu, India, spanning from 2011 to 2023. Dr.P. Selvakumar's academic and research interests lie primarily in the field of phytochemistry. He has made significant contributions to his field, with an impressive record that includes 30 journal publications, 12 patents, and 20 conference papers. Additionally, he has organized 5 seminars and conferences, further demonstrating his commitment to academic and scientific exchange. Throughout his career, Dr. P.Selvakumar has been recognized for his dedication and excellence in teaching, having been honored with several awards, including the Best Teacher Awards. He holds a prominent position in the academic community, contributing actively to research.

Santhikumar Rajamahanthi has been working as Professor of Mathematics and Dean of Student Affairs of Basic Science and Humanities in Aditya Institute of Technology and Management, K.Kotturu, Tekkali, Srikakulam Dist, Andhra Pradesh, India. He received M.Sc (Mathematics) from Osmania University, Hyderabad. He awarded Ph.D. from Acharya Nagarjuna University, Gunturu, Andhra Pradesh. He published 36 research papers in various International Journals and presented many papers in different International Conferences. He filed 3 patents. He authored 2 books and 1 book Chapter. He has been acting as a reviewer for SCI journals and Editor for a reputed International Journal. He has been acted as Convener, Reviewer, Resource person and Judge in many National/International Workshops/Conferences. He acted as BOS member of a state university. He successfully completed UGC-SERO project. He received C V Raman Prize 2022 from Institute of Researchers, Kerala. He is life member of ISTE, IMS, ISPS, AP&TSMS, InRes. His area of research is Relativity and Cosmology. He has been actively involving in Research, Academic and Professional development activities for the last 22 years.

Lakshmana Rao Agatamudi has been working as Professor of Statistics and Associate Dean of Academics of Basic Science and Humanities in Aditya Institute of Technology and Management, K.Kotturu, Tekkali, Srikakulam Dist, Andhra Pradesh, India. He served as Associate Professor of Statistics, Department of Statistics, College of Natural Science, Jimma University, Ethiopia from 2015 to 2019. He received M.Sc (Statistics) from Andhra University, Visakhapatnam.

He awarded Ph.D. degree in Statistics from Andhra University, Visakhapatnam. He published 25 research papers in various International Journals and presented 9 papers in different International Conferences. He has 3 patents. He is a reviewer for Scopus journals and Editor for a reputed International Journal. He has been acted as Convener, Reviewer, Resource person and Judge in many International Conferences. He successfully completed UGC-SERO project. He is life member of ISTE, ORSI, ISPS. He is recipient of award Best oral presentation of 2013 from Karunya University, Coimbatour, Tamilnadu. His area of research is Inventory model, Queuing models and Distribution models. He has been actively involving in Research, Academic and Professional development activities for the last 27 years.

Sourabh Singh Roy is a distinguished academic associated with the Faculty of Management Studies at Gopal Narayan Singh University, located in Sasaram, Bihar. With a strong commitment to excellence in management education, Dr. Roy contributes significantly to the academic and professional development of students, blending practical insights with theoretical knowledge to prepare future leaders in the field.

S. Haashni is a student at Vimal Jyothi Convent Matriculation Higher Secondary School in Coimbatore, Tamil Nadu, India. She is passionate about playing tennis and enjoys spending her free time reading books. Haashni is known for her enthusiasm for learning and her active involvement in both academics and extracurricular activities.

Atharva Saraf is pursuing his undergraduate studies in Computer Science at the School of Engineering, Ajeenkya DY Patil University. His research interests are cybersecurity, artificial intelligence, and data science. He has published book chapters in reputed national/international publishers such as CRC Press and IGI Global, and presented his research results at various conferences.

Ashwarya Sharma is an accomplished academic and researcher in the field of commerce, currently serving as an Assistant Professor at Chandigarh Group of Colleges, Jhanjeri, Mohali since 2016. She holds an M.Com (Hons) in Finance, Banking, and Insurance from Panjab University and a B.Ed from Punjabi University, Patiala. She is also pursuing a Ph.D. from Punjabi University with a focus on corporate governance and social responsibility. With a strong background in research, Ashwarya has presented and published papers at national conferences and in international journals. Her academic contributions are complemented by certifications from the National Stock Exchange of India and multiple NCFM modules. Recognized for her dedication, she has received accolades such as "Employee of the Month," "Women Employee of the Year," and a "Certificate of

Best Performance." Beyond academics, she actively volunteers for street animal welfare and has participated in environmental initiative

Mohit Sharma is a research scholar at the Institute of Management Studies and Research, Maharshi Dayanand University, Rohtak, Haryana, India. His research focuses on the dynamic relationship between brand image and purchase intentions, particularly in the context of brand extensions within the FMCG sector. He has co-authored several scholarly works and continues to contribute to the academic community. Mohit can be reached at mohitattri.rs.imsar@mdurohtak.ac.in

Koppala Venugopal is a Professor of MBA and currently serves as the Associate Dean (R&D), Vice-President (IIC), and Convenor (NISP) at Aditya Institute of Technology and Management. With over 25 years of academic and research experience, he has successfully completed six major research projects funded by UGC, UOG and other organizations. Dr. Venugopal has contributed significantly to the field, authoring 121 research papers, publishing 10 books, and presenting 86 papers. His extensive work also includes advisory and editorial roles in 13 publications, 4 consultancy projects, 2 corporate Trainings, 56 invited talks, 2 granted patents, and the supervision of 6 research scholars.

Index

M

N

O

P

R

S

T

V

www.ingramcontent.com/pod-product-compliance
Ingram Content Group UK Ltd.
Pitfield, Milton Keynes, MK11 3LW, UK
UKHW050946290926
15470UKWH00005B/1667

9 798369 399347